KU-743-024

Cheeney
1980 May.

Dimensional Methods in Engineering and Physics

Dimensional Methods in Engineering and Physics

reference sets and the possibilities of their extension

E. de St Q. Isaacson

and

M. de St Q. Isaacson

Edward Arnold

©1975 E. de St Q. Isaacson and M. de St Q. Isaacson

First published 1975
by Edward Arnold (Publishers) Ltd
25 Hill Street, London W1X 8LL

ISBN 0 7131 3847 3

All Rights Reserved. No part of this publication
may be reproduced, stored in a retrieval
system, or transmitted in any form or by any
means, electronic, mechanical, photocopying,
recording or otherwise, without the prior
permission of Edward Arnold (Publishers) Ltd.

Text set in 10/12 pt. IBM Press Roman, printed by photolithography,
and bound in Great Britain at The Pitman Press, Bath

To Rachel
and
to Lorraine-Ann

Preface

The simple dimensional argument that shows the period of a pendulum to be proportional to $(l/g)^{1/2}$ is well known. That results of this nature can readily be obtained without recourse to fundamental analysis is liable to impress the student as strange and rather wonderful. Indeed, dimensional analysis is a by-way of physics that seldom fails to fascinate even the hardened practitioner.

The appeal, however, lies in the physical applications of the theory rather than in the underlying mathematics. An effort, therefore, has been made to keep mathematics in this book to a minimum and we have not hesitated, for example, to use standard results taken without proof from the algebraic theory of linear dependence. Those who are unfamiliar with the background will find no difficulty in referring to any of the numerous texts available. For a similar reason the Buckingham pi-theorem has been developed from semi-intuitive appeals to physical principles. In this case, however, we have thought it advisable to support the argument by a more rigorous mathematical treatment provided as an appendix.

We believe that the book will be of theoretical interest and practical use, not only to the student, but to those working professionally in the general fields that are considered. The level of difficulty is such that it will be fully grasped by the first-year undergraduate reading engineering or physics but, in order to assist with a first approach, certain passages are contained within the symbols ● . . . ●. These passages may require more application than is necessary for the bulk of the text and may be omitted if required.

While much of the text is based on standard material, a number of new results have been included and certain items do not appear to have been previously treated in the literature from the viewpoint that we have ourselves adopted. This remark applies particularly to our discussion of 'physical independence' and its application to the possibilities of increasing the precision of results by working with definitions made in terms of reference sets extended beyond the classical mass–length–time triad. The relationship of 'physical independence' to the legitimacy of model distortion is also considered.

It is impossible to discuss dimensions without treating with units. This is accepted, but our concern remains essentially dimensional; that is, we are interested in the manner in which units may be combined rather than with the magnitude of the units operated upon. We relegate to the background considerations of, say, metres and feet, concentrating rather on the underlying concept of the quantity 'length'.

A selection of problems will be found in Appendix 2. These are not regarded as an integral part of the text. Our intention, rather, has been to provide the reader with an assortment of intellectual exercises of varying levels of difficulty and involving the application of some of the techniques which are discussed.

Finally, we have endeavoured to keep abreast with recent developments and we include results based upon or referring to the latest available literature.

London and Cambridge   E. de St Q. I.
1975   M. de St Q. I.

Contents

List of More Frequently Used Symbols

A	amplitude	g	unspecified function
	area	H	dimension of heat
a	acceleration	h	heat-transfer coefficient
	velocity of sound		height
B	magnetic flux density		Planck's constant
C	capacitance	I	moment of inertia
	couple		dimension of electric
C_D	drag coefficient		current
C_f	skin-friction coefficient	i	electric current
C_L	lift coefficient	J	current density
C_M	moment coefficient	K	bulk modulus
C_p	pressure coefficient	k	undetermined constant
c	compressibility		scale factor
	velocity of light		radius of gyration
c_p	specific heat at constant		roughness parameter
	pressure		Boltzmann's constant
c_V	specific heat at constant	L	dimension of length
	volume		inductance
D	drag force		lift force
d	diameter	l	length
	distance	lg	log to base 10
E	energy	ln	log to base e
e	charge on electron	M	dimension of mass
e	base of natural logarithms		moment
F	dimension of force		Mach number
Fr	Froude number	m	mass
f	undetermined (implicit)		magnetic pole strength
	function	N	Nusselt number
f	force	n	frequency
G	gravitational constant		number of variables
Gr	Grashof number	P	power
g	initial acceleration in	Pr	Prandtl number
	free fall	p	pressure

Q	dimension of electric charge	x	
	flow rate (volume/time)	y	cartesian co-ordinates
q	electric charge	z	'unknown quantities'
	quantity of heat	α	angle
R	radial length dimension		coefficient of thermal
	resistance		expansion (linear)
	gas constant	β	coefficient of thermal
$\mathfrak{R}$	universal gas constant		expansion (vol.)
r	rank of indicial matrix	γ	ratio of specific heats (c_p/c_V)
	number of reference dimensions	ϵ	strain
	radius		permittivity
Re	Reynolds number	η	efficiency
			modulus of rigidity
S	Strouhal number	Θ	dimension of temperature
St	Stanton number	θ	temperature
s	distance		angle
	entropy	κ	thermal conductivity
T	dimension of time	λ	wave length
t	time		mean free path
	tension	μ	dynamic viscosity
			permeability
U	characteristic velocity	ν	kinematic viscosity
	internal energy		Poisson's ratio
u	velocity at a point	π	dimensionless product
	initial velocity	ρ	resistivity
u^*	velocity characterising		density
	turbulence		radius of curvature
V	potential difference:	σ	normal stress
	voltage		electrical conductivity
v	velocity (terminal)	τ	surface tension
	volume		shear stress
W	work	ϕ	undetermined (explicit)
	weight		function
Wb	Weber number		scalar potential function
		Ψ	dimension of length component
X			(tangential)
Y	component length dimensions	ω	angular velocity
Z			angular frequency

ASRM absolute significance of relative magnitude
DP   dimensionless product

PP   power product

[1]   'of zero dimensions'

$\equiv$ 'is dimensionally equivalent to . . .'

a the vector 'a'

Where units are used, conventional symbols will apply: e.g. m = metres, s = seconds, lbf = pound force, etc.

Occasional departures from or extensions to the symbolism listed above will be specifically mentioned in the text.

1

Units of Measurement and Physical Dimensions

1.1 Units and measurement

In order to render a physical concept quantitative rather than qualitative, that concept must be defined in such a way that it becomes capable of measurement. This is implied by linguistic usage when we speak of a physical 'quantity'.

Where we have no guidance as to how the measurement of a concept is to be carried out, then that concept cannot be regarded as belonging to physics. Compare the idea of viscosity, as measured in terms of poises, with the pre-scientific notion of 'imperfect fluid mobility': compare the idea of temperature, as measured in degrees Celsius, with the pre-scientific approach involving feelings of warmth and coolness based on physiological sensation.

Our use of the term 'quantity', then, will be confined to concepts which are measurable by some well-defined process. (The words 'quantity' and 'variable' will be largely interchangeable, but we shall prefer the latter when we wish to emphasise that the magnitude of a quantity occurring in some physical situation is related to and varies with the magnitudes of other quantities.)

The measurement of a quantity generally involves a *unit* of measurement. Examples are the metre, erg, farad, light-year, tonne per hour, mile per gallon, lumen per square foot, etc. It will be clear that nomenclature is no guide to complexity and the erg is not necessarily a simpler concept than is the mile per gallon.

We introduce a definition: *a unit is a selected magnitude of a physical quantity in terms of which other magnitudes of the same quantity may be expressed as multiples*. Thus, a length of 6.3 centimetres is equal to 6.3 times the unit length of one centimetre.

Measurement *may*, but does not necessarily, entail the use of units. The sine of an acute angle could be used as a measure of its size: the logarithm of a number could be used as a measure of its magnitude. Other less naive examples will be considered in **1.4**. Nevertheless, the unit plays a fundamental role in physics and, for reasons to be made clear, most physical measurements are, in fact, carried out in terms of units.

1.2   Relationships between units

We could conceivably allocate an isolated and independent unit to each physical quantity, basing this unit on a reference standard. Mass, length and time may, for example, be measured in terms of the kilogram, the metre and the second and there is no theoretical reason why the magnitudes of other quantities should not also be based on standard units. But a gain in simplicity and, more important, an increase in our knowledge of the interrelatedness of the concepts of physics are made possible as soon as 'complex' quantities and their units are defined in terms of simpler ones.

Thus, a unit of area, the square metre, is related to the metre, a unit of length, rather than to a standard unit of area. Similarly, a unit of work, the joule, is related to the newton and the metre — units of force and length — rather than to any standard unit of work. And, provided the newton and the metre are themselves well defined, we require no standard unit quantity of work to act as our reference.

In this fashion the units of complex or derived quantities are related to simpler ones and, in laying bare this relationship we are introducing a structure into our physical thinking which is essential if any worthwhile development of physical theory is to take place.

1.3   Reference units

This process, however, will clearly have a limit. If we are to escape circularity, we must decide upon some set of 'reference' quantities, the units of which are not to be defined in terms of simpler ones. Many writers denote such quantities as 'fundamental' or 'basic', but we shall avoid this usage as the quantities chosen do not differ in any philosophical sense from the other quantities which may be derived from them.

The selection of reference quantities is, then, largely a matter of convention and our choice is liable to be influenced by psychological determinants. We instinctively feel that length and *time* are more 'elementary' quantities than are, say, length and *velocity*, although we could perfectly readily regard time as being derived from reference units based upon length and velocity. Indeed, Asimov[1] * suggests that a convenient unit for the measurement of small passages of time would be the 'light-metre', representing the time taken for light to traverse a metre length *in vacuo*. And, from another viewpoint it might be difficult to demonstrate logically that even such 'obviously' derived quantities as torque or viscosity were, indeed, less simple in character than, say, mass.

The set of reference quantities most commonly and conventionally selected

---

* References will be found listed on page 213.

are mass, length and time. And, in principle, we shall find that all other mechanical quantities are such that their units may be defined in terms of the units of this set (1.7). A practical factor influencing the choice is the ease and accuracy with which these three quantities may be measured, for mass, length and time may each be determined with considerably greater precision and convenience than can, say, area or energy. (Compare the practical difficulties involved in measuring the surface area of a lump of sugar with the ease of finding its mass.)

In particular, the allocation of stable standard units to mass, length and time is a problem of no great technical difficulty. Thus, before the introduction of the *Système International d'Unités* (the SI), the standard unit of *mass* was the International Prototype Kilogramme, a solid cylinder of diameter equal to its height and made of platinum—iridium (90 Pt : 10 Ir). This unit was established in 1889 and is preserved at the Bureau International des Poids et Mesures, Sèvres, France.

The standard unit of *length* was the International Prototype Metre, a bar of the same alloy and having a special X cross-section devised by Tresca to provide maximum rigidity. The metre was defined as the distance between two transverse graduation lines when the bar was maintained at 0°C and supported on two rollers in a horizontal plane at two positions 0.0571 metres apart.*

The standard unit of *time* was the Mean Solar Second, defined as 1/86 400 of the mean solar day, a particularly unsatisfactory standard in view of the pronounced irregularities in the rate of the earth's rotation.

With the introduction of the SI, a considerable gain in precision was effected, and the basic units of mass, length and time (together with those of temperature, electric current and luminous intensity) are currently defined as follows.

Mass: The unit of mass called the *kilogram* is the mass of the international prototype which is in the custody of the Bureau International des Poids et Mesures (BIPM) at Sèvres. (3rd Conférence Générale des Poids et Mesures (CGPM), 1901)

Length: The unit of length is called the *metre* and is 1 650 763.73 wavelengths *in vacuo* of the radiation corresponding to the transition between the energy levels $2p_{10}$ and $5d_5$ of the krypton-86 atom. (11th CGPM, 1960)

Time: The unit of time called the *second* is the duration of 9 192 631 770 cycles of the radiation corresponding to the transition between the two hyperfine levels of the fundamental state of the caesium-133 atom. (13th CGPM, 1967)

Thermodynamic temperature: The unit of thermodynamic temperature called the *kelvin* is the fraction 1/273.16 of the thermodynamic temperature of the triple point of water. (13th CGPM, 1967)

* There is no real element of circularity in this definition!

Electric current: The unit of electric current called the *ampere* is that constant current which, if maintained in two parallel rectilinear conductors of infinite length, of negligible circular cross-section, and placed 1 metre apart in a vacuum, would produce between these conductors a force equal to 2×10^{-7} newton per metre length. (9th CGPM, 1948)

Luminous intensity: The unit of luminous intensity called the *candela* is the luminous intensity, in the perpendicular direction, of a surface of $1/600\,000$ square metres of a black body at the freezing temperature of platinum under a pressure of 101 325 newtons per metre squared. (13th CGPM, 1967)

1.4   The absolute significance of relative magnitude

A variety of units may conveniently be used to determine the magnitude of the same physical quantity. Length may be measured in microns, fathoms or light-years: work may be measured in foot-pounds, ergs or joules. It is, however, desirable that the relative magnitude of two concrete examples of the same quantity should be independent of the units used. Thus if the density of acetamide, when measured in grams per millilitre, is 1.43 times that of aniline, we should expect this relative density to remain the same if both measurements be made in pounds per cubic foot. Again, if the velocity of compression waves in aluminium, when measured in metres per second, is 2.67 times the velocity in lead, we should expect this relative velocity to remain unaltered if both measurements be made in miles per hour.

This is the requirement which Bridgman[5] calls the principle of the 'absolute significance of relative magnitude' (ASRM). We are interested in determining the nature of the restriction that has to be placed upon the units used in order that this requirement may be met, and we shall show that it is necessary and sufficient that the units of derived quantities be formed from the products of the powers of the units of those quantities taken as comprising the reference set. We use the term 'power product' (PP) to indicate this type of association. Illustrative of a simple case is the (derived) unit of acceleration, the 'metre per second per second' which is expressible in the form $(\text{metre}) \times (\text{second})^{-2}$, a product of powers of the reference units of length and time.

● *Using an argument due to Bridgman[5], we first prove that if relative magnitudes are to have absolute significance, then derived units must necessarily be measured in terms of PPs. Let A, B and C represent the reference quantities in terms of which a derived quantity S is to be measured. (We assume three reference

---

* Passages contained between the symbols ● . . . ● may conveniently be omitted on a first reading.

quantities only, but the argument is general and may be extended to a set containing any number of members.) We have, then, $S = f(A, B, C)$, where f is an arbitrary function save only that it is taken as continuous and differentiable.

Now suppose we make two specific measurements involving the derived quantity and, in consequence, we obtain the values S_1 and S_2. The corresponding magnitudes of the reference units will be respectively $A_1, B_1 \, C_1$ and A_2, B_2, C_2. The requirement of ASRM now entails that

$$\frac{S_1}{S_2} = \frac{f(A_1, B_1, C_1)}{f(A_2, B_2, C_2)}$$

must remain constant for all changes in scale of the units in which A, B and C are measured.

Let these units be so changed that they assume $1/x$, $1/y$ and $1/z$ of their previous values. Then the measures of the reference quantities in the two examples considered will now be xA_1, yB_1, zC_1 and xA_2, yB_2, zC_2. It follows that

$$\frac{f(A_1, B_1, C_1)}{f(A_2, B_2, C_2)} = \frac{f(xA_1, yB_1, zC_1)}{f(xA_2, yB_2, zC_2)}$$

Our aim is to determine the nature of the unknown function f and we proceed by writing

$$f(xA_1, yB_1, zC_1) = f(xA_2, yB_2, zC_2) \cdot \frac{f(A_1, B_1, C_1)}{f(A_2, B_2, C_2)}$$

Differentiate partially with respect to x and let f' denote the partial derivative of f with regard to the arguments xA_1 or xA_2 as the case may be. Then, since

$$\frac{\partial f}{\partial x} = \frac{\partial f}{\partial (xA_1)} \cdot \frac{\partial (xA_1)}{\partial x} = A_1 f'$$

we have

$$A_1 f'(xA_1, yB_1, zC_1) = A_2 f'(xA_2, yB_2, zC_2) \cdot \frac{f(A_1, B_1, C_1)}{f(A_2, B_2, C_2)}$$

Let us put $x = y = z = 1$ to obtain

$$A_1 \frac{f'(A_1, B_1, C_1)}{f(A_1, B_1, C_1)} = A_2 \frac{f'(A_2, B_2, C_2)}{f(A_2, B_2, C_2)}$$

which is certainly true for all values of A_1, B_1, C_1 and A_2, B_2, C_2.

If we now hold A_2, B_2, C_2 constant, while permitting A_1, B_1, C_1 to vary, we may remove the subscripts and write the preceding relationship in the form

$$\frac{A}{f} \cdot \frac{\partial f}{\partial A} = \text{constant} = a \text{ (say)}$$

(Here we have written $\partial f/\partial A$ for f', which is justifiable since $f' = \partial f/\partial (xA)$ and x are being held equal to 1.) Integration gives $f = k_1 A^a$, where k_1 is a function of B and C only.

The same process may now be repeated, differentiating partially with respect to y and to z in succession and integrating. The final result will be

$$f = S = k \cdot A^a B^b C^c$$

where a, b, c, k are constants. Hence the derived quantity is necessarily expressible as a product of the powers of the reference units — multiplied, possibly, by a numerical constant.

So much for necessity. It remains for us to show that it is sufficient for a derived quantity to be expressed as a PP for ASRM to be maintained. The proof here is more straightforward.

With the notation already established, consider the ratio of the values of two specific measurements of a derived quantity

$$\frac{S_1}{S_2} = \frac{A_1{}^a B_1{}^b C_1{}^c}{A_2{}^a B_2{}^b C_2{}^c}$$

Now if the units of measurement be changed to $1/x$, $1/y$ and $1/z$ of their original values, we shall have as our ratio

$$\frac{S_1}{S_2} = \frac{(xA_1)^a (yB_1)^b (zC_1)^c}{(xA_2)^a (yB_2)^b (zC_2)^c} = \frac{A_1{}^a B_1{}^b C_1{}^c}{A_2{}^a B_2{}^b C_2{}^c} \cdot \frac{x^a y^b z^c}{x^a y^b z^c}$$

The right-hand factor is equal to unity and constancy of the ratio is therefore maintained. ●

The requirement of ASRM is of basic importance and it is, indeed, to satisfy this principle that physical quantities are generally measured and defined in terms involving PPs. (Measurements of the reference quantities themselves will, of course, satisfy ASRM since the measurement concerned constitutes a (degenerate) PP, the reference unit entering the PP with exponent equal to one and the other reference quantities entering with exponent zero.)

Where quantities are not so measured and, in particular, where measurements are not made in terms of units as defined in 1.1, then the proof just considered will be invalid and ASRM will not hold good. The ratio of two specific measurements made under such conditions will not generally be absolute.

We consider a few examples. The 'hardness' of a material may be determined by the Brinell test (involving measurement of the impression made by a steel ball pressed into the surface) or it may be measured by the scleroscope test (involving a determination of the height of the rebound of a steel ball dropped onto the surface). Now if two materials have a Brinell hardness of, say, 460 and 170 respectively, they will have equivalent hardnesses as shown by the scleroscope of 63 and 26. The ratios of hardness will, in consequence, be 2.71 or 2.42 according to which means is adopted for measurement. We have a vague feeling that, since no well-defined units are used, such measurements are 'not properly scientific'.

Again, ASRM does not apply if the magnitude of a quantity is measured by ranking, as when the geologist resorts to Moh's scale to determine the hardness of a mineral. This merely involves the observation of whether it will scratch or be scratched by a number of graded standard specimens.

Similarly, ASRM fails if measurements are taken from different datum levels — as in the Celsius and Fahrenheit temperature scales. (Compare the ratio of $10°C$ to $5°C$ with the ratio of the same temperatures expressed in the Fahrenheit scale, that is $50°F$ and $41°F$.) Note, however, that ASRM necessarily applies as soon as we consider two temperature *ranges*, since the arbitrary datum level drops out and the degree Celsius and the degree Fahrenheit become true units, in that each temperature range is expressible in terms of multiples of the degree.

Measurements of sound-level differences in terms of decibels have been cited as an exception to the rule that units must be in the form of PPs if ASRM is to hold. Two sounds having root-mean-square sound-pressure levels of p_1 and p_2 will differ by n decibels where $n = 20 \lg (p_1/p_2)$. n, then, is not measured in terms of PPs, yet the ratios of two decibel measurements will, nevertheless, be independent of the units in which the p's are measured. The exception, however, is apparent only, for the decibel is not a unit of sound and the absolute magnitude of a sound — as opposed to the ratio between two sounds — cannot be expressed in decibels. When a sound is loosely stated to have a magnitude of n decibels, it is generally being expressed as a ratio to a reference sound level of $0.000\ 02\ N/m^2$, which approximates to the threshold of the human ear.

This apparently artificial approach to measurement is occasionally used where intensities are judged by a subjective physiological response.* Thus we might, if we so wished, introduce a measure, W, of the sensation of weight, which could be related to the physical concept of weight by some such relationship as $W = \lg(w/w_0)$, w_0 being a reference weight. W would not then represent a measurement of magnitude in terms of units and ASRM could not be applied. A more 'respectable' instance occurs in astronomy, where the difference in the observed magnitudes of two stars is given by

$$m_1 - m_2 = -2.5 \lg (E_1/E_2)$$

---

* Cf. Weber's law relating sensation to stimulus.

where E_1 and E_2 are the stellar luminances expressed in lux. There are many other examples of measurements not based on unit magnitudes and to which ASRM does not apply (e.g. pH values, colours and wind strengths). But we would not wish to place too great an emphasis on these exceptions, for most physical quantities are, in fact, deliberately defined in such a way as to ensure that ASRM is valid and they are, therefore, necessarily expressible in terms of products of the powers of the reference units.

1.5   Dimensions

That a derived unit may be expressed as a PP suggests that the unit is possessed of a structure and we now take up this question of structure in more detail. Following Clerk Maxwell[47] we introduce the term 'dimension' and we define the dimensions of a physical quantity in terms of the 'code' that informs us of the manner of the derivation of the units in which that quantity is measured, the code, that is, which indicates the structure of the unit as a particular PP. More specifically, Maxwell chooses mass, length and time as his reference units and defines the dimensions of a unit by the statement:

> When a given unit varies as the nth power of one of these reference* units, it is said to be of n dimensions as regards that unit.

If then a quantity, q, has the dimensions of a, b and c respectively with regard to mass, length and time, we say briefly that its dimensional representation is given by $q \equiv M^a L^b T^c$.

We refer indifferently to the dimensions of a *quantity* or the dimensions of one of the *units* of that quantity, and we now consider examples showing how these dimensions may be derived and how they follow logically from the definition of the quantity concerned. Note that a change of scale from one unit to another of the same quantity will not effect the dimensions. (For example, the dimensions of the quantity 'work' will be the same, no matter whether it be reckoned in joules, in ergs or in foot pound-forces.)

Area: The unit of area is defined as (unit of length)^2 and we say immediately that the dimensions of the quantity area (or of any unit of area) are 2 with regard to length, since the derived unit of area varies as the 2nd power of the reference unit of length. The dimensional representation of area, then, is given by $A \equiv L^2$.

More generally, area may be defined as the integral of length with respect to a second length measured orthogonally to the first, that is $A = \int y \, . \, dx$. Since this integral represents the limit of the sum of a sequence of products of two lengths,

---

* We have substituted the word 'reference' for the word 'fundamental' which occurs in the original text.

each of the form $y \,.\, \Delta x$, it has the same dimensions as $y \,.\, x$ and again we have $A \equiv L^2$.

By an extension of this approach, we say that the integral $\int\int \ldots \int w \,.\, dx \,.\, dy \ldots dz$ has the same dimensions as $w \,.\, x \,.\, y \ldots z$. And, as a special case of this last result we see that the dimensions of *volume* are given by $v \equiv L^3$.

Velocity is defined as the derivative of distance with respect to time, that is $v = ds/dt$. This differential is the limit of the fraction $\Delta s/\Delta t$. The dimension of an increment of length, Δs, is L, while that of Δt is T. It follows that the unit of velocity varies as the 1st power of L and as the (-1)th power of T. We then write the dimensions of velocity as L/T or LT^{-1}.

Similarly *acceleration*, defined as $d^2 s/dt^2$ has the dimensions of $\Delta(\Delta s/\Delta t)/\Delta t$, or LT^{-2}, and, more generally $d^n s/dt^n$ will have the dimensions of LT^{-n}.

Force is defined by the equation $f = ma$. Writing M as the dimension of mass and recalling that a has the dimensions LT^{-2}, it follows that the dimensions of force are given by $f \equiv MLT^{-2}$.

Viscosity provides a less straightforward illustration. The defining equation of this quantity is

$$\mu = \frac{\text{shear stress in a fluid}}{\text{velocity gradient normal to flow}}$$

We then argue progressively:

dimensions of force $\equiv MLT^{-2}$ (as already determined)

dimensions of stress $\equiv$ dimensions of (force/area)

$$\equiv MLT^{-2}/L^2 \equiv ML^{-1}T^{-2}$$

dimensions of
velocity gradient $\equiv$ dimensions of (velocity/distance)

$$\equiv LT^{-1}/L \equiv T^{-1}$$

From which it follows immediately that

dimensions of viscosity, $\mu \equiv MLT^{-2}/T^{-1} \equiv ML^{-1}T^{-1}$

By similar reasoning, based upon the defining equation, we may determine the dimensions of any other physical quantity, provided only that this be measured in units which are of the structure of a PP.

In the foregoing discussion we have taken as our reference set the quantities mass, length and time. We now make the possibly obvious point that the dimensions of a quantity are dependent upon the quantities chosen for inclusion in this set. If, for instance, we had taken force ($\equiv F$) as a reference quantity in place of mass ($\equiv M$), then the dimensions of viscosity would have been $FL^{-2}T$, since when working with F, L and T the dimensions of stress reduce simply to FL^{-2}. The argument set forth above would then have been modified accordingly.

Again, had we chosen F, L and T as our reference set, the quantity 'mass' would have been regarded as derived and its dimensions would be represented by $m \equiv FL^{-1}T^2$, as becomes clear by consideration of the equation $m = f/a$.

This confirms that there is nothing 'absolute' about the dimensional representation of a quantity, since this is dependent upon the reference set chosen, nor is there any deep distinction between 'derived' and 'reference' quantities, for, subject to limitations to be discussed in 1.7, any arbitrary quantity may be selected for reference purposes. We shall ourselves generally follow the convention that opts for a reference set based on mass, length and time; the dimensions of some of the commoner physical quantities that may be derived from M, L and T will be found listed in Table 1 of Appendix 1.

A dimensional symbol which occurs in both numerator and denominator may be 'cancelled out', an operation which has already been tacitly performed during the course of our discussion. This procedure may be justified as follows: Suppose that a reference quantity A enters into the numerator of the dimensional representation of a derived quantity q with exponent n and into the denominator of that representation with exponent m. Considering other reference units held constant, we have for changes in magnitude of the unit of A, $q \propto A^n/A^m$ or $q \propto A^{n-m}$. The dimensional symbol A^m has, been effectively 'cancelled out'.

From much the same point of view, the cancellation of units, including derived units, is commonplace in elementary arithmetic. If a car moving at 30 miles/hour consumes petrol at the rate of 20 miles/gallon, the number of hours it runs on one gallon will be

$$20 \, \frac{\text{miles}}{\text{gallons}} \div 30 \, \frac{\text{miles}}{\text{hours}} = \frac{2}{3} \, \frac{\text{hours}}{\text{gallons}}$$

Here the unit of length, the mile, has been cancelled out.

Such examples make it clear that an algebraic symbol representing a quantity in a physical equation is never to be interpreted as a number but rather as a number of *specific units* each of the appropriate dimension. Consider the equation $s = ut + \frac{1}{2}gt^2$, giving the distance fallen by a body in time t. If we substitute the values of $t = 2$ seconds, $u = 1.0$ metres/second and $g = 9.80$ metres/second^2, we emphatically do not obtain the result that $s = 2.16$. We find, rather that $s = 21.6$ *metres*, for

$$s = 1.0 \, \frac{\text{metres}}{\text{seconds}} \times 2 \text{ seconds} + \frac{1}{2} \times 9.80 \, \frac{\text{metres}}{(\text{seconds})^2} \times (2 \text{ seconds})^2$$

$$= 21.6 \text{ metres}$$

Here the units (seconds) and (seconds)^2 have respectively been cancelled in the two terms on the right-hand side of the equation.

1.6 Pure numbers, dimensional and dimensionless constants, and variables

The dimensions of a physical quantity may reduce to zero. The quantity is then referred to as 'dimensionless' or as a 'numeric' and has the nature of a number.

Dimensionless quantities may consist simply of the ratios of two similar quantities. Thus strain, defined as 'change in length divided by original length', has the dimensions of $dl/l \equiv L/L \equiv [1]$, where [1] symbolises the dimensionless aspect of a numeric. Further examples are angle and relative humidity. More generally, a dimensionless quantity or numeric may consist of any combination of physical quantities such that each of the reference dimensions M, L and T reduces to zero in the representative dimensional formula.

Such dimensionless combinations of quantities are known as 'dimensionless products' or DPs. Extensive use of DPs will be made in the following pages and we mention a few examples. A typical DP is the Reynolds number, defined as $Re = vl\rho/\mu$ where

v, a velocity, has dimensions $\quad LT^{-1}$

l, a length, has dimensions $\quad L$

ρ, a density, has dimensions $\quad ML^{-3}$

μ, a dynamic viscosity has
        dimensions $\quad ML^{-1}T^{-1}$

It follows that the dimensions of Re will be given by

$$Re \equiv \frac{LT^{-1}.L.ML^{-3}}{ML^{-1}T^{-1}} \equiv M^0 L^0 T^0 \equiv [1]$$

Further instances of simple DPs, selected at random are $t^2 g/l$, $v^2/(lg)$ and $E/(v^2 \rho l^3)$ where, in the last case, the quantity E is energy and ρ is density. The reader is recommended to check the dimensionless nature of these (unnamed) products for himself. The Mach number, $M = U/a$, where both U and a are velocities, is a well-known example of a (named) DP based on a ratio.

Since in a DP each reference quantity cancels out between the numerator and the denominator, it follows that the magnitude of the DP will be independent of the magnitude of the reference units used. Thus if the value of the Reynolds number relating to a certain situation is 5000 when the component variables are measured in SI units, then it will remain unchanged at that figure if those variables be measured in, say, the foot—pound—second system.

The terms appearing in a physical equation may conveniently be regarded as involving pure numbers standing alone or in association with dimensional and dimensionless variables and constants. Examples of these possibilities are listed below:

1. *Pure numbers* are those that arise out of mathematical operations and are not directly based upon physical quantities, for example

$$\frac{4}{3} \quad \text{in} \quad v = \frac{4}{3}\pi r^3$$

$$e \quad \text{in} \quad N = N_0\, e^{-\lambda t}$$

2. *Dimensionless constants* are 'numeric' constants that derive from physical quantities, for example

π (defined as a ratio of lengths)

1838 (the ratio of the mass of a hydrogen atom to the rest mass of an electron)

3. *Dimensionless variables* such as strain, Reynolds number and Poisson's ratio.

4. *Dimensional constants* such as the velocity of light, initial acceleration in free fall and Planck's constant.

5. *Dimensional variables* such as mass, torque and viscosity.

We proceed with a few comments. Pure numbers arise as a result of mathematical manipulations performed upon the original defining equations of the quantities involved or upon the equations of motion of the system considered. They tend, therefore, to be small (e.g. e or $\sqrt{2}$) and are often 'neat' integers or rationals. Bond[3] has discussed this matter statistically. Some 1500 values were collected by random selection from different sources and their frequency distribution was investigated. This showed empirically that the probability of the magnitude of a numerical coefficient falling outside the range $1/n$ to n is simply $1/n$, which, of course, falls off rapidly with increasing n.

This is not to deny that very large (or small) numbers may arise in pure mathematics. A famous example, mentioned by Littlewood[44], is the Skewe's number, defined as the lowest value of x such that the number of primes $\leqslant x$ exceeds $\int_0^x dx/\ln x$, and which is approximately equal to $10^{10^{10^{34}}}$. This, however, is in no sense a numerical coefficient arising in a physical equation as a result of routine manipulations.

Where numerical constants arise directly from a defining relationship, the relevant units are generally selected in such a manner as to ensure that the value of any constant of proportionality becomes equal to unity. (We chose our units in order to be able to write $v = ds/dt$ rather than, say, $v = (2/3)ds/dt$, but see also 8.4.)

We do not insist upon any fundamental distinction between pure numbers and dimensionless constants. We often tend, for example, to regard π as a pure number rather than as a DP, even though its definition is based upon a ratio of lengths.

Dimensional constants, as opposed to pure numbers, arise from empirical situations and are liable to be clumsy. They may often be very large (e.g. the velocity of light, Young's modulus, Avogadro's number, etc.) or very small (e.g. the charge on an electron, Planck's constant, mass of a hydrogen atom, etc.). Use of this observation will be made in 11.2 and a more detailed discussion of dimensional constants will be found in 3.3.

In contrast with the case of DPs, the value of all dimensional quantities will depend upon the magnitude of the units used in the measurement of the reference quantities. If, for instance, in the foot–pound–second system, Young's modulus of a specimen of granite is $E = 7.0 \times 10^6$ lbf/in^2, then in SI units it will be $E = 4.82 \times 10^{10}$ N/m^2. To effect such conversions we may most simply consider the structure of the units involved. In the present instance we have only to write

$$\frac{\text{lbf}}{\text{in}^2} = \left(\frac{\text{lbf}}{\text{N}} \times \frac{\text{m}^2}{\text{in}^2}\right) \cdot \frac{\text{N}}{\text{m}^2}$$

and, since 1 lbf = 4.45 N and 1 in = 0.0254 m, we have immediately

$$1\frac{\text{lbf}}{\text{in}^2} = \left(4.45 \times \frac{1}{0.0254^2}\right) \frac{\text{N}}{\text{m}^2} = 6890\frac{\text{N}}{\text{m}^2}$$

1.7   Comments upon MLT as a reference set

We consider the questions:

1. Are the dimensions M, L and T sufficient to describe the structure of the units of all physical quantities?

2. How valid are our reasons for the adoption of M, L and T as members of the reference set?

With regard to the sufficiency of what we now refer to as the MLT system, we observe that the whole structure of classical mechanics may be developed by taking mass, length and time as undefined quantities and by progressively defining all other quantities directly or indirectly in terms of them. We may, for example, proceed by developing such concepts as velocity, force and work, defined in terms of the equations $v = ds/dt, f = ma$ and $W = \int f \cdot ds$. We then continue in this fashion to develop those further quantities required to describe the wide variety of phenomena which we encounter. It is, then, a matter of actual observation that no quantity is in fact introduced into mechanics that cannot be derived from MLT, and it follows that M, L and T are wholly sufficient for the purposes of dimensional representation.

They are not, however, *necessary*, for, as has already been seen, we could with equal logic develop a consistent theory of mechanics on the basis of alternative sets of undefined reference quantities. There is, indeed, another dimensional system which still finds favour and which is based upon the quantities force, length and time (the FLT system).

As soon as we leave the equations of simple mechanics and introduce thermal, electrical or magnetic quantities, it may be useful to introduce reference sets that extend beyond MLT (or FLT). This will not generally be *necessary*, for there may frequently be a relationship between these new quantities and such well-established quantities as mechanical energy which are themselves susceptible of representation in terms of MLT. Nonetheless, there will often be excellent heuristic reasons for working with extended sets, and we shall see in Chapter 5 that it may be advantageous to work with extensions to the basic reference set of three members even when treating with problems lying within the restricted field of mechanics itself.

We pass now to our second question. Having shown that the choice of MLT as a reference set is neither necessary nor capable of theoretical justification, we make the point that the selection of a reference set is, nevertheless, not to be made in a wholly arbitrary fashion. It is, for example, clear that we are liable to land in difficulties if we chose length, time and velocity; in such a system, we should be unable to define quantities involving mass and there would, moreover, be no unique manner of expressing quantities involving velocity.

It is desirable, then, that each unit of the reference set be independent in the sense that it is not possible to express it as a combination of the others. Intuitively this is because a unit which can be derived from the other members of the set contributes nothing new to the set and may, therefore, be dispensed with.

This requirement is not always obvious of fulfillment. Consider a system in which the proposed set of reference dimensions consists of velocity V, impulse I, and energy E. These are related to the MLT system by the following equivalences:

$$V \equiv M^0 L^1 T^{-1}$$

$$I \equiv M^1 L^1 T^{-1}$$

$$E \equiv M^1 L^2 T^{-2}$$

If we attempt to treat mass, length and time as derived quantities and to express them in terms of the proposed new reference units, we shall land in an impasse, for we find that M is indeterminate and that there is no solution for L and T. The reason is that E may itself be derived from V and I by virtue of the relationship $E = VI$, and it follows that E is not an independent quantity and contributes nothing to the set.

Algebraically the condition of independence is as follows. Suppose a proposed set of quantities P, Q and R can be expressed in terms of MLT as:

$$P = M^a L^b T^c$$

$$Q = M^d L^e T^f$$

$$R = M^g L^h T^i$$

Then, P, Q and R will each be independent of the others if and only if the determinant

$$\begin{vmatrix} a & b & c \\ d & e & f \\ g & h & i \end{vmatrix} \neq 0$$

Subject to this condition, P, Q and R may be used as a set of reference quantities. (For the algebraic background to this argument see Littlewood[43].) With this in mind we now see that, in the example involving V, I and E, the relevant determinant is

$$\begin{vmatrix} 0 & 1 & -1 \\ 1 & 1 & -1 \\ 1 & 2 & -2 \end{vmatrix}$$

which reduces to zero. Hence arises the inconsistency that we noticed.

We introduce a definition. If, from a number of quantities, a set be chosen such that

1. no member of the set may be derived from (or is dependent upon) any other member or members of the set; and

2. each quantity not included in the set may be so derived;

then the set is referred to as a 'complete' set. This is an important concept and one which will have a variety of applications.

What has been said concerning the sufficiency of MLT as a three-member set of reference dimensions applies equally, then, to any other set of three quantities that may be put forward as the basis of a dimensional system, provided only that the proposed set be complete in the sense of the foregoing definition.

It will, however, be noted that, while we have shown explicitly how to check the suitability of three quantities as members of a reference set by expressing those quantities in terms of MLT, we have, nevertheless, been able to check the adequacy of M, L and T themselves only by observing empirically that their use leads to no inconsistency. (Notwithstanding this remark, we shall see **4.3** that, although conditions for the completeness of the reference set MLT are generally satisfied, there may occur specially restrictive situations, involving a limited number of variables, in which their mutual independence breaks down. This need not, however, concern us at the present stage.) It is clear, then, that at least one triad of reference quantities must be tested empirically; once the adequacy of that set has been demonstrated, any other set may subsequently be referred to it and tested by an examination of the relevant determinant.

1.8   The geometrical approach

Relationships deriving from the dimensional structures of quantities may be illustrated by means of directed graphs. An example based upon a few simple quantities is shown in figure 1, which will be largely self-explanatory. The positive or negative integer placed at the sides of the directed 'edges' refers to the power to which the quantity at the initial point of the edge is raised when it enters into the definition of the further quantity at which the edge terminates.

Figure 1 Relationships between quantities

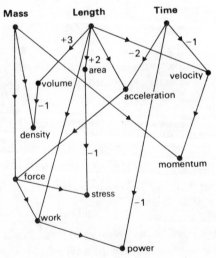

The value of +1 is understood where no other indication is given.

The graph relating to a given set of quantities is not uniquely defined and, although some workers such as Happ[33] have made use of this approach in their treatment of certain problems, it appears that a deeper insight into the relationships between quantities may be derived from the use of a more sophisticated technique which considers a 'dimension space' apparently first discussed by Corrsin[10].

We consider a set of three reference dimensions and, to fix our ideas, we chose MLT. We proceed to establish an orthogonal framework, measuring exponents of M, L and T along each of the three axes. Any physical quantity may then be represented by a unique point in the space defined by this framework. Thus velocity, with the dimensions LT^{-1}, will correspond to the point v as shown in figure 2 or, alternatively, to the vector directed to that point from the origin of the framework.

Figure 2

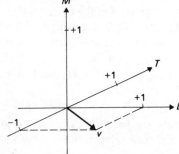

The reader may now readily satisfy himself that:

1. The multiplication of two or more quantities results in a vector which is the sum of the vectors representing the individual quantities.

2. The kth power of any quantity is represented by the product of the vectorial representation of that quantity with the scalar k.

3. A DP regarded as a product of zero dimensions, is represented by a sequence of vectors, corresponding to each of the quantities contributing to that DP. This sequence, moreover, starts and terminates at the origin, thus forming a 'cycle'.

The relationship $v^2 = sg$ is shown represented in figure 3, which is based on a framework consisting of the two dimensions L and T. Should we wish to represent the DP sgv^{-2}, we have only to reverse the sense of the vector v^2; that

Figure 3

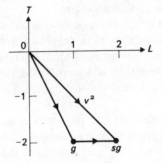

is, we multiply it by the scalar -1 to obtain $(v^2)^{-1}$ and we note that the DP corresponds, as expected, to a cycle.

A number of results may be readily demonstrated by using this approach. Thus, we shall see in 7.1 that the Reynolds number can be expressed as a ratio of forces. It is, however, already clear that all DPs may be expressed as a ratio of similar quantities. To prove this we argue as follows. Let the cycle representing a DP consist of the points $O\,A_1\,A_2\,\ldots A_n\,O$, where O is the origin of the framework. Select any point A_k on this cycle. Consider the two quantities represented by the vector sums $OA_1 + A_1A_2\,\ldots + A_{k-1}A_k$ and $A_kA_{k+1} + \ldots + A_nO$. These two quantities will be equivalent to the vectors OA_k and A_kO, that is to the ratio of two identical quantities.

● We shall not explore such applications further. We revert, rather, more specifically to the discussion in 1.7 and now make the point that any three quantities may be used as a reference set, provided that their representative vectors are independent, that is provided that they do not lie in the same plane. This approach may be readily extended to encompass the case of a reference set of n (> 3) members, represented by an orthogonal cartesian system in n-space. Here the condition of independence of n quantities is that their representative vectors do not lie in any space of $(n - 1)$ dimensions (the word 'dimensions' here being used in the sense of conventional (analytical geometry). ●

2

The Basic Idea of Dimensional Analysis

2.1 The combining of physical quantities

In physics, unlike quantities cannot be added to, subtracted from or equated with one another, nor can they be regarded as greater or less than one another. Accelerations may be added to accelerations, but never to viscosities or strains. 3 inches plus 4 seconds equals 7 of nothing at all, and 1 tonne is neither greater nor less than 1 metre. Different units of the same quantity may, however, be added. 3 inches plus 4 centimetres is a quantity with the dimension of length; problems of conversion will not concern us.

Although in physics unlike quantities may not be added, they may be associated by multiplication or division. Having glossed over this possibility in the last chapter, we now consider it in more detail. We may, for instance, 'divide' 10 oranges by 5 boys to obtain 2 'oranges per boy' or, symbolically, 2 oranges/boy. Similarly, 10 men each working for 5 hours constitutes, by an exercise in 'multiplication', 10 x 5 'man hours'.

The processes of 'division' and 'multiplication' as applied to physical quantities or their units differ from the processes by which pure members are divided or multiplied, for, in the arithmetical sense of the words, to divide or multiply a 'distance' by a 'time' is without meaning. But in the physical sense we find that, when two quantities and their respective magnitudes are divided or multiplied, then the numbers associated with the units are operated upon arithmetically, while the units themselves are combined dimensionally to form the relevant PP.

Let us clarify this. When we say that 6 metres divided by 3 seconds is equal to to 2 m/s, it is evident that the numbers 6 and 3 have wholly lost their identity as a result of the arithmetical operation which throws up the number 2. But the derived quantity of 'velocity', formed by 'dividing' metres by seconds, is expressed in m/s units which have the dimensions LT^{-1}. These units, moreover, are such that the units of length and time are only loosely associated and continue to maintain their original identity.

The name of a derived unit, such as a foot pound or a kilowatt hour may occasionally give a clue to its structure — but not necessarily, as may be seen by considering the joule. All these units are, of course, measures of work and have the same structure as indicated by their dimensional representation of $ML^2 T^{-2}$.

The process of dividing or multiplying units, then, is symbolic and operational. It obeys well-understood formal rules and finds its justification in the fact that the magnitude of the derived quantity is required to vary directly or inversely with the magnitudes of the simpler quantities. These simpler quantities are accordingly associated with one another by 'multiplication' or 'division' respectively (cf. 1.5). Other forms of association of physical quantities, by exponentiation for example, are not generally employed, for the resultant quantities would not then fulfill the requirement of ASRM.

2.2 Dimensional homogeneity

Since the terms of a physical equation are liable to be added to, subtracted from or equated with one another, it follows that each term will represent a like quantity and that the terms will, therefore, be dimensionally equivalent to one another. As an example we return to the equation $s = ut + \frac{1}{2}gt^2$ and notice that the dimensional representation of each of the three terms reduces to L, that is to a distance. Indeed we have:

s = total *distance* travelled by body $\equiv L$

ut = *distance* travelled in time t, had the body moved with constant
        velocity u $\equiv (L/T) . T \equiv L$

$\frac{1}{2}gt^2$ = additional *distance* travelled in time t, as a result of acceleration g
        $\equiv (L/T^2) . T^2 \equiv L$

This situation is typical of any 'properly constructed' physical equation, that is any equation that results from mathematical operations performed upon basic defining relationships. Such equations are said to be characterised by 'dimensional homogeneity'. (In the literature, a dimensionally homogeneous equation is often referred to as a 'complete' equation, but we shall not ourselves follow this usage.)

Occasionally, however, we encounter an equation which lacks dimensional homogeneity even though it be descriptive of a physical situation. When this happens, the equation concerned will generally be an empirical one which merely describes the results of an observation and which has not been mathematically developed from theoretical principles. Examples and a further discussion will be found in 3.4.

Alternatively, as a more trivial exception, we have the 'bastard' equation formed by a haphazard algebraic combination of equations relating to different quantities. To quote Porter[55]:

It is undoubtedly possible to add the numerical values of an ounce and a day to a foot but the result will have no physical meaning. It is, moreover, true that we can write down an equation such as

$$s + v = \tfrac{1}{2}gt^2 + gt$$

for a body falling from rest and this equation can be satisfied at every moment. Such an equation, however, never appears in the course of a physical investigation. It is in reality two equations with their terms intermingled and each of these equations is separately satisfied.

The cases of 'empirical' and 'bastard' equations, then, represent no significant exceptions to the general rule, and we are able to state categorically that, provided an equation has been theoretically derived, its terms will inevitably be dimensionally homogeneous. This principle is of basic importance and we proceed to demonstrate its validity by a consideration of the manner in which physical equations are actually constructed.

2.3 The construction of physical equations

Although no formal proof of the dimensional homogeneity of physical equations appears available, it represents, nevertheless, a special case of a very general principle in analysis. Thus Jaques Monod[49] is considering the whole field of scientific enquiry when he tells us that 'It is in fact impossible to analyse any phenomenon in terms other than those of the invariants that are conserved through it'.

We shall ourselves attempt to justify this principle by showing how the equation-building process depends essentially upon the discovery of different aspects of the same quantity and the writing of them down as equal.

In mechanics we equate potential energy lost with kinetic energy gained; in electrical network theory we equate currents arriving at a junction with those leaving that junction; in problems dealing with heat we equate heat arriving at a surface by conduction with heat leaving that surface by radiation. To quote Duncan[15], an equation

> may be the symbolic representation of the fact that all components of a force acting upon a body in a particular direction, including the 'inertia forces' are in balance or it may express the fact that the net mass of fluid entering a fixed region per unit of time is equal to the increase of the contained mass in the unit of time.

That is, forces are equated with forces, masses with masses, and so on.

It follows that, while the several terms of a physical equation may appear superficially different, their dimensional structure must necessarily be the same. We proceed to illustrate this by the deduction of two specific equations.

2.3.1   Atmospheric pressure as a function of height

The equation giving atmospheric pressure in terms of height above sea level is derived as follows. In an element of unit cross-section and of thickness Δx, the change in pressure $-\Delta p$ between the lower and upper boundaries of the element (see figure 4) is equal to the weight of air that is contained in the element. We

Figure 4

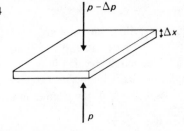

therefore *equate two expressions for the change in pressure*:

$$-\Delta p = \rho g \,.\, \Delta x$$

Now, assuming constant temperature, we may take the density ρ as proportional to the pressure. In our equation, then, we may substitute kp for ρ and, taking to the limit, we have

$$\frac{dp}{dx} = -kpg \tag{1}$$

which may be integrated to give the required result:

$$p = p_0 e^{-kgx}$$

But the substitution $\rho = kp$ needs clarification. It may be objected that we here equate a 'density' with a 'pressure', which is hardly in conformity with what has been said. Since this is typical of a common procedure, we examine more closely what has actually been done. The statement that 'density is proportional to pressure' implies that if ρ_0 and p_0 represent one pair of corresponding values of density and pressure, and if ρ_1 and p_1 represent any other pair of values, then

$$\frac{\rho_0}{\rho_1} = \frac{p_0}{p_1}$$

that is, *one dimensionless ratio is equal to another*. This yields $\rho_1 = (\rho_0/p_0)p_1 = k \,.\, p_1$. It follows that k is not a pure number but a dimensional constant with the same dimensions as ρ/p. The substitution $\rho = kp$, therefore, *equates two expressions for a density*, which is fully in accordance with the requirement of dimensional homogeneity. The effect of k as a multiplier is, then, dual. It changes the numerical value of the right-hand side of equation 1 in order to maintain numerical equality; it also changes the dimensions of the right-hand side in order to maintain homogeneity.

Note too that when, in our mathematical manipulation, we come to integrate $\int dp/p$, the result is not $\ln p$. This would be meaningless, since we can take logarithms only of numbers of numerics and not of dimensional quantities such as pressures (3.2). The result is rather ($\ln p$ + constant of integration) which is equivalent to $\ln (p/p_0)$, that is the logarithm of a ratio.

2.3.2 Bernoulli's theorem

As our further example we derive Bernoulli's theorem for fluid flow. This more detailed illustration of the equation-forming process shows clearly the roles played by:

a) a qualitative understanding of the nature of the phenomenon under discussion,

b) the equating of quantities of similar dimensions, and

c) mathematical manipulation.

These three stages are of quite general application.

Figure 5

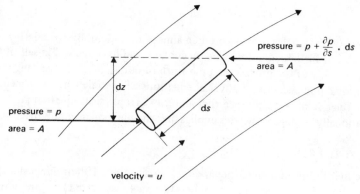

The fluid is taken as incompressible and inviscid and the flow steady. Consider, as in figure 5, a slender cylindrical element of length ds and cross-sectional area A, the axis of which is colinear with a streamline. We proceed to examine the various forces contributing to the motion of the fluid. These are:

1. The external force acting on the fluid in the streamline direction. Since the fluid is inviscid and hence cannot sustain shear stresses, this force will be due only to the difference in pressure acting at the two ends of the cylinder, that is $-(\partial p/\partial s) \cdot ds \cdot A$.

2. The component of the gravitational force in the free streamline direction, that is $-\rho A g (\partial z/\partial s) \cdot ds$.

The mass of the fluid element is $\rho A \cdot ds$ and its acceleration will be $u \cdot (\partial u/\partial s)$. Now the mass—acceleration product will clearly be a force, as follows from the defining equation $f = ma$, and for steady flow this force will derive from and be equal to the sum of the two forces previously considered. We therefore *equate the net forces along the streamline* to obtain

$$\rho A u \frac{\partial u}{\partial s} \cdot ds = -\frac{\partial p}{\partial s} \cdot ds \cdot A - \rho g A \cdot ds \cdot \frac{\partial z}{\partial s}$$

Straightforward mathematical manipulation now gives

$$\frac{1}{\rho} \frac{\partial p}{\partial s} + u \frac{\partial u}{\partial s} + g \frac{\partial z}{\partial s} = 0$$

in which each term has the dimensions LT^{-2}, and then finally

$$\frac{p}{\rho} + \tfrac{1}{2} u^2 + gz = \text{constant} \tag{2}$$

These two examples have been taken almost at random. By considering the deduction of further equations, the reader will readily convince himself of the general validity of the process, and the truth of our assertion concerning the dimensional homogeneity of physical equations will be taken as confirmed.

One final point. We have seen that all terms in a physical equation are dimensionally homogeneous; this implies that the equation may be written in the form

$$a_1 A_1 + a_2 A_2 + \ldots a_m A_m = 0 \tag{3}$$

where the A's are equi-dimensional quantities (possibly DPs or dimensional constants) and the a's are numerical coefficients (possibly ± 1) or functions of one or more DP.

Thus equations 1 and 2 just considered may be written in the form of equation 3 as

$$1 \cdot p - e^{-kgx} \cdot p_0 = 0$$

and $\quad 1 \cdot (p/\rho) + \tfrac{1}{2} \cdot u^2 + 1 \cdot gz - 1 \cdot \text{constant} = 0$

Let us now divide equation 3 throughout by A_m to obtain

$$a_1 \frac{A_1}{Am} + a_2 \frac{A_2}{Am} + \ldots + a_m = 0 \tag{4}$$

which is clearly a relationship between DPs only and numerical coefficients.

We have already seen (1.6) that the magnitude of a DP remains unchanged as a result of changes in scale of the reference units, and it follows that the magnitude of any numerical coefficient in a dimensionally homogeneous equation will also be unchanged. To quote a simple illustration: if in SI units we have $t = 2\pi\sqrt{(l/g)}$ (which may be put in the form of equation 3 as $t^2 g - 4\pi^2.l = 0$), then the coefficient 2π will remain unchanged if the variables in the equation be measured in the foot—pound—second or in any other system.

2.4 Buckingham's pi-theorem

We come now to the central theorem of dimensional analysis, first explicitly stated by Buckingham[8] in 1914 and known as the Buckingham 'pi-theorem'. We express this in the form:

> If there is a dimensionally homogeneous equation relating n quantities defined in terms of r reference dimensions, then the equation may be reduced to a relationship between $(n - r)$ independent DPs provided that the members of the reference set be themselves so chosen as to be independent of one another.

We shall point out in Chapter 4 that the $(n - r)$ DPs form a complete set of DPs and we may, accordingly, rephrase Buckingham's theorem more concisely in the alternative form:

> If an equation be dimensionally homogeneous, it may be reduced to a relationship between a complete set of DPs.

Note that we use the term 'complete' here in the sense of 1.7. A set of DPs formed from n quantities is complete if no member of the set may be derived from the other members and if all other DPs based on the same n quantities may be so derived.

Illustrating Buckingham's pi-theorem with the equation $s = ut + \frac{1}{2} gt^2$, we divide throughout by s to obtain

$$1 = (ut/s) + \tfrac{1}{2} (gt^2/s) \tag{1}$$

Each of the terms is now dimensionless, and we have very simply reduced the original equation to a relationship between the two DPs (ut/s) and (gt^2/s).

Two points should be noted here. Since in the original equation there were $n = 4$ variables and since these are defined in terms of $r = 2$ independent reference quantities only, namely L and T, we have, as expected, $(n - r) = (4 - 2) = 2$ DPs in our set. Secondly we mention that the criteria of completeness are satisfied by this set of two DPs in view of the fact that:

1. neither DP may be derived from the other, since the quantities u and g each appear in only one member of the set;

2. any other DP formed from the quantities s, u, g and t may be derived from the two members of the set. (That this is indeed so follows from an argument given in **4.3**.)

For illustrative purposes we give examples of two further DPs based upon the quantities s, u, g and t and which may be expressed in terms of the set of two DPs thrown up by equation 1:

1. (u/gt) which equals $(ut/s) \times (gt^2/s)^{-1}$, and

2. (u^2/sg) which equals $(ut/s)^2 \times (gt^2/s)^{-1}$

Note that these two further DPs themselves form a complete set, which illustrates that such sets are not uniquely defined.

We should perhaps mention that the situation may be concisely described in the language of group theory and, in fact, a complete set of DPs, together with their inverses (reciprocals), constitute a set of independent generators of an Abelian group under the operation of multiplication.

A formal algebraic proof of Buckingham's pi-theorem is set out in the appendix to this chapter (**2.6**). Alternative proofs may be based upon the geometrical approach of **1.8** due to Corrsin[10] and upon the approach involving matrices discussed in **4.3** and **4.4**.

Meanwhile we content ourselves with a less rigorous indication of the result. We have already shown that any dimensionally homogeneous equation may be expressed in the form of equation 4 of **2.3** and that this represents a relationship between DPs rather than between dimensional variables. It remains, then, to make the further point that the number of separate DPs occurring is $(n - r)$. To see this we rewrite equation 3 of **2.3** as an unspecified function of n variables:

$$g(a_1, a_2, \ldots, a_n) = 0 \tag{2}$$

And we rewrite equation 4 of the same section as an unspecified function of k variables (DPs):

$$f(\pi_1, \pi_2, \ldots, \pi_k) = 0 \tag{3}$$

It is clear that the number of variables in equation 3 has been reduced from the number appearing in equation 2 as a result of their combining to form DPs. This combination, moreover, has been possible only as a result of $r = 3$ functional restrictions placed upon the new variables, regarded as DPs, by virtue of the fact that the exponents of M, L and T will each be made zero. For any reasonably well-behaved functions g and f, it follows intuitively, though the formal proof is unwieldy, that the original number of variables, n, may then be reduced by r to yield $k = (n - r)$ DPs in the final result. This argument makes clear the desirability of the requirement that the reference dimensions should be independent; failing this, the number of restrictions placed upon the original variables will be less than r and the number of DPs produced will, in consequence, be greater than $(n - r)$.

We have still to show that the set of $(n - r)$ DPs is complete. The proof of this may most readily be carried out in terms of the theory of linear equations and, while we shall not be setting out full details of the mathematical treatment, an appropriate reference will be found in **4.3** and an intuitive argument will be provided at the end of **4.4**.

This, then, is Buckingham's pi theorem. It represents a severe and necessary restriction which must be imposed upon the equations of physics for, as we have seen, all such equations will be of a structure that makes possible their reduction to the form of equation 3. It is in this observation that the germ of dimensional analysis lies.

2.5 An example of dimensional analysis: the oscillations of a simple pendulum

A classical example, by which the student is frequently introduced to dimensional analysis, considers the oscillations of a simple pendulum. We treat this problem in some detail, as it will provide useful orientation for what is to follow.

From our physical insight into the nature of the situation we suspect that there will be an equation based upon the quantities:

t period of oscillation
l length of string
g acceleration due to gravity
α angle of swing.

This equation may be represented as $f(t, l, g, \alpha) = 0$ and will correspond to equation 2 of the last section.

Our initial aim must be to seek information concerning the reduced equation, written in terms of DPs and corresponding to equation 3. We have, therefore, to enquire how the four quantities listed above may be combined into dimensionless forms, and this will clearly necessitate a preliminary examination of the dimensional structure of the units in which the quantities are measured. We find, in fact, that

$$t \equiv T, \quad l \equiv L, \quad g \equiv LT^{-2} \quad \text{and} \quad \alpha \equiv [1]$$

Note that the dimensions of α follow from the definition of angle as a ratio of lengths, that is 'arc/radius'.

Now, since there are 4 variables, and since we are working with the two reference dimensions L and T, it follows that there will be $(n - r) = (4 - 2) = 2$ DPs in a complete set. We notice that $(t^2 g/l)$ and (α) are two independent DPs which may be formed from the variables, and we suspect that these do, in fact, comprise the required set.*

Buckingham's pi-theorem now tells us that any relationship between the four original variables may be reduced to a relationship between these DPs. We have then

$$f(\pi_1, \pi_2) = f(t^2 g/l, \alpha) = 0$$

Solving explicitly for $(t^2 g/l)$ gives

$$t^2 g/l = \phi(\alpha) \tag{1}$$

or $\quad t = \sqrt{(l/g)} \cdot \phi(\alpha)$

where ϕ is some undetermined function of α only.

At this stage we have made an important deduction: we have shown that for constant α the period of oscillation t must vary as $\sqrt{(l/g)}$. But as long as our reasoning is based solely on considerations of measurement and dimension we can obtain no information concerning the dependence of the period of oscillation upon its amplitude α. Detailed conventional analysis, however, shows that

$$\phi(\alpha) = 2\pi \left[1 + (\tfrac{1}{2}k)^2 + \left(\frac{1.3}{2.4} k^2 \right)^2 + \left(\frac{1.3.5}{2.4.6} k^3 \right)^2 + \ldots \right]$$

where $k = \sin \tfrac{1}{2}\alpha$. Provided that α is sufficiently small for second and higher powers of k to be neglected, this series reduces to $\phi(\alpha) = 2\pi$, corresponding to the elementary formula $t = 2\pi\sqrt{(l/g)}$.

Had we restricted our enquiry to the situation in which α is small, and had our physical intuition been such that we realised that under such conditions α was no longer a significant variable, it would have been possible to argue directly that the only DP to be obtained from the three quantities t, l and g is $(t^2 g/l)$. It would follow in this case that the complete set of DPs contains one member only and that the relationship connecting the variables is necessarily of the

---

* Systematic methods for the construction of complete sets of DPs will be given in
  Chapter 4.

form

$$f(t^2 g/l) = 0$$

Since the value of f is here constantly equal to zero, it is implied that the argument of the function, that is $(t^2 g/l)$, must itself be constant and, in consequence,

$t^2 g/l$ = constant

or $t = k\sqrt{(l/g)}$

which is the required relationship.

Where, as in this case, dimensional analysis yields a solution in which there are no undetermined functions but only an undetermined constant, we say that we have obtained a 'complete solution'.

Note that had we erred by assuming that the mass m of the pendulum bob was a significant variable affecting the period of oscillation, we should have found that no DP containing m could be formed, since none of the other variables is defined in terms of mass. It follows that our analysis would then have shown that m does not and cannot enter into the problem; or, more explicitly, that there is no dimensionally homogeneous equation that can be based upon the variables t, l, g and m.

This, incidentally, is precisely what we would expect in the light of Buckingham's theorem. By introducing the quantity m of dimension M, we are increasing the number of variables from n to $(n + 1)$ and at the same time increasing the number of reference dimensions from r to $(r + 1)$. The number of DPs, being equal to $(n - r)$, is therefore left unchanged.

We have in this chapter operated somewhat uncritically. Our purpose, however, has been to illustrate the basic thinking which underlies the idea of dimensional analysis. Rigour will be added and subtleties discussed in the following pages.

● 2.6   Appendix: proof of Buckingham's pi-theorem

We now provide a formal proof of Buckingham's pi-theorem, based upon an approach given by Bridgman[5]. The theorem will be proved in the form:

If there is a dimensionally homogeneous equation relating n quantities defined in terms of r reference dimensions, then the equation may be reduced to a relationship between $(n - r)$ independent DPs, provided that the members of the reference set are themselves so chosen as to be independent of one another.

Denote the variables (quantities) by $a_1, a_2, \ldots, a_n$ and let them satisfy a dimensionally homogeneous functional equation

$$f(a_1, a_2, \ldots, a_n) = 0 \tag{1}$$

(Here the a's are used indistinguishably to represent the quantities themselves and their numerical magnitudes.)

For the sake of simplicity, we proceed with our argument on the assumption that $r = 3$ and that the reference dimensions are MLT, but the reader will readily satisfy himself that the proof may be generalised for any value of r. Express $a_1, a_2, \ldots, a_n$ as PPs based upon the reference dimensions, thus obtaining

$$a_1 = M^{r_1} L^{s_1} T^{t_1}$$
$$a_2 = M^{r_2} L^{s_2} T^{t_2}$$

.

.

.

$$a_n = M^{r_n} L^{s_n} T^{t_n}$$

Now decrease the size of the units in which the reference quantities MLT are measured by factors x, y and z respectively. This will result in the numerical magnitudes of $a_1, a_2, \ldots, a_n$ becoming

$$a'_1 = (x^{r_1} . y^{s_1} . z^{t_1}) a_1$$
$$a'_2 = (x^{r_2} . y^{s_2} . z^{t_2}) a_2$$

.

.

.

$$a'_n = (x^{r_n} . y^{s_n} . z^{t_n}) a_n$$

Since equation 1 is dimensionally homogeneous, it will, as we saw in 2.3, hold good for any change in the units of the reference quantities and we may, therefore, put

$$f(a'_1, a'_2, \ldots, a'_n)$$
$$= f(x^{r_1} y^{s_1} z^{t_1} a_1, x^{r_2} y^{s_2} z^{t_2} a_2, \ldots, x^{r_n} y^{s_n} z^{t_n} a_n) = 0$$

Differentiate this last equation partially with regard to x to obtain

$$r_1 x^{r_1-1} y^{s_1} z^{t_1} a_1 \frac{\partial f}{\partial a'_1} + r_2 x^{r_2-1} y^{s_2} z^{t_2} a_2 \frac{\partial f}{\partial a'_2} + \ldots$$

$$+ r_n x^{r_n-1} y^{r_n} z^{t_n} a_n \frac{\partial f}{\partial a'_n} = 0 \qquad (2)$$

Now, we are entitled to choose any values we wish for x, y and z and we select the case where $x = y = z = 1$. In consequence we have $a'_i = a_i$ for all i's, and equation 2 reduces simply to

$$r_1 a_1 \frac{\partial f}{\partial a_1} + r_2 a_2 \frac{\partial f}{\partial a_2} + \ldots + r_n a_n \frac{\partial f}{\partial a_n} = 0 \qquad (3)$$

Select now a set of new variables $A_1, A_2, \ldots, A_n$ defined in such a way that

$$A_i = a_i^{1/r_i} \qquad \text{or, alternatively,} \qquad a_i = A_i^{r_i}$$

This definition entails that the A's will be of dimension 1 with regard to M. We also have

$$\mathrm{d}a_i = r_i \frac{a_i}{A_i} \cdot \mathrm{d}A_i$$

and, in consequence,

$$r_i a_i \frac{\partial f}{\partial a_i} = A_i \frac{\partial f}{\partial A_i}$$

Substituting this in equation 3 gives

$$A_1 \frac{\partial f}{\partial A_1} + A_2 \frac{\partial f}{\partial A_2} + \ldots + A_n \frac{\partial f}{\partial A_n} = 0 \qquad (4)$$

The number of variables in equation 4 may now be reduced from n to $(n-1)$ by dividing throughout by A_n. We accordingly define a set of new variables B_1, $B_2, \ldots, B_{n-1}$ in terms of the ratio

$$B_i = \frac{A_i}{A_n}$$

and have, in consequence;

$$f(A_1, A_2, \ldots, A_n) = f(A_n B_1, A_n B_2, \ldots, A_n B_{n-1}, A_n)$$

Note that the function on the right-hand side of this last equation is evidently independent of A_n, as follows from the fact that its derivative with regard to

A_n vanishes. Indeed,

$$\frac{\partial f}{\partial A_n} = B_1 \frac{\partial f}{\partial A_n B_1} + B_2 \frac{\partial f}{\partial A_n B_2} + \ldots$$

$$\ldots + B_{n-1} \frac{\partial f}{\partial A_n B_{n-1}} + \frac{\partial f}{\partial A_n}$$

$$= [A_1 (\partial f / \partial A_1) + A_2 (\partial f / \partial A_2) + \ldots + A_n (\partial f / \partial A_n)] / A_n$$

which is zero, since the numerator is equal to the left-hand side of equation 4.

It follows that $f(A_n B_1, A_n B_2, \ldots, A_n B_{n-1})$ is a function only of the $(n-1)$ B's, and we may write

$$F(A_1, A_2, \ldots, A_n) = f(A_n B_1, A_n B_2, \ldots, A_n B_{n-1}) = g(B_1, B_2, \ldots, B_{n-1})$$

Note further that since all the A's are, as we have seen, of dimension 1 in M, the B's, being defined as ratios of the A's, will themselves be dimensionless in M.

Repeat now the argument from the start, putting $g(B_1, B_2, \ldots, B_{n-1}) = 0$, which follows from equation 1, since the value of g is identically equal to that of the function f. But g = 0 is an equation of the same type as equation 1, with the difference that one variable has disappeared from the argument and one reference quantity has no longer any place in the definition of the variables.

Repetition of the process and involving the differentiation of g with respect to y will again reduce the number of variables in the argument by one and the new variables will, moreover, be dimensionless in both M and L.

One final and similar repetition involving differentiation with respect to z will reduce the number of variables to $(n - r) = (n - 3)$ only, and these, moreover, will be dimensionless in each of the three reference quantities MLT. Furthermore, since all changes in variable involve only raising to a power or taking a ratio, the final $(n - 3)$ variables will be dimensionless products of powers, that is, they will be DPs formed from the original variables $a_1, a_2, \ldots, a_n$. This proves Buckingham's pi-theorem.

Two additional points. We have shown that the number of DPs is $(n - r)$; that these also comprise a complete set follows from a consideration of the 'indicial matrix' to be discussed in Chapter 4. Secondly, we have proved that there are $(n - r)$ DPs in the case where the r reference dimensions are independent. The perceptive reader will have noticed that where this is not the case and where, for instance, there is some linear relationship between r_i, s_i and t_i which holds good for all i's, then the proof breaks down. We shall show, again in Chapter 4, that the difference between the number of the original variables, n, and the number of DPs is, in fact, equal to the number of *independent* reference dimensions, this latter number possibly varying with the situation considered. ●

3

The Dimensional Homogeneity of Physical Equations

3.1   The terms of a physical equation

When we say that the 'terms' of a properly constructed equation are dimensionally homogeneous, we define a 'term' as a group of symbols associated with one another by multiplication and/or division and separated from other terms by signs of addition, subtraction or equality. Thus the left-hand side of equation 2 of 2.3.2 contains three terms. The equation as a whole contains four terms.

A term may itself contain subsidiary terms. An expression such as $a \cdot \lg(b_1 + b_2)$ may be taken as a single term. The b's, being linked additively within the main term, will be regarded as subsidiary terms and will themselves be homogeneous since, were they not so, their addition or subtraction would have no physical meaning.

No particular difficulties are liable to arise, but we further illustrate the position with the equation giving the loss of kinetic energy that follows the collision of two imperfectly elastic bodies. This is

$$\Delta KE = \tfrac{1}{2} \frac{m_1 m_2}{m_1 + m_2} (v_1 - v_2)^2 (1 - e)$$

e, being a coefficient of restitution, is a ratio and, therefore, dimensionless. The single main term on the right-hand side has the dimensions of energy, that is ML^2T^{-2}. The subsidiary terms, linked by minus signs within the brackets, are each homogeneous, those in the first bracket having the dimensions of LT^{-1} and those in the second bracket having the dimensions [1]. This example shows that the distinction between main and subsidiary terms is convenient rather than rigid, for the equation under consideration may readily be multiplied out to produce six main terms on the right-hand side, each with the dimensions of energy. Again, if we multiply throughout by the denominator $(m_1 + m_2)$ there will result a total of eight main terms in the equation, two on the left and six on the right, each, this time, with the dimensions of $M^2L^2T^{-2}$.

3.2   Occurrence of DPs as the arguments of functions

The argument of a logarithmic, trigonometrical, exponential or any transcendental function must necessarily be dimensionless. It is, for example, meaningless to talk of the logarithm of an electric charge of the sine of a time. There follow a few examples illustrating this principle:

$$W = p_1 v_1 \ln (v_2/v_1)$$
(1)

(Giving the work done in changing the volume of a gas isothermally. The argument of the logarithm, a ratio of volumes, is dimensionless.)

$$y = A \sin \omega(t - x/v)$$
(2)

(Giving the displacement due to a progressive wave, A being the amplitude, $\omega/2\pi$ the frequency, t the time, v the velocity of progression and x the distance of the point considered from the origin. The argument of the sine function is dimensionless.)

$$y = \frac{dF}{dx} = \frac{1}{\sigma\sqrt{(2\pi)}} e^{-(x - \mu)^2/2\sigma^2}$$
(3)

(Giving the change in frequency of a normally distributed quantity x about its mean value μ. Here σ, the standard deviation has, by definition, the same dimension as x (and, therefore as μ). The exponent is, therefore, dimensionless.)

To quote Buckingham[8]:

Such expressions as log Q or sin Q do not occur in physical equations, for no purely arithmetical operator, except a simple numerical multiplier, can be applied to an operand which is not itself a dimensionless number, because we cannot assign any definite meaning to the result of such an operation.

That a theoretically derived equation is expressible in a dimensionally homogeneous form does not mean, however, that the homogeneity may not be masked by mathematical manipulation. Starting with $v = s/t$ we might uncritically write: $\lg v = \lg s - \lg t$. In this form, however, the equation is meaningless and would not normally be used to advance any physical argument.

Less obvious is an equation quoted by Bridgman[5]. With a slight change in notation, this may be written

$$\lg C = -\frac{\lambda}{R\theta} + \frac{a}{R}\lg\theta + \frac{b}{R}\theta + \frac{c}{2R}\theta^2 + A$$

Here C is a concentration of gas, λ is a heat, θ is a temperature, a, b and c are dimensional coefficients (with a/R dimensionless) and A is a constant of integration. An inspection of this equation shows that a rearrangement of the terms is possible, and we may group together lg C, (a/R) log θ and A into the single term

lg $(C/B\theta a/R)$

where B is a new constant with dimensions equivalent to those of $C \theta^{-a/R}$. We now have a dimensionally homogeneous equation and the argument of the logarithm is dimensionless, as is to be expected.

That the argument of any transcendental function is dimensionless may be neatly demonstrated as follows. Let the function be

$$\phi(\lambda) = \sum_{i=-\infty}^{\infty} \alpha_i \lambda^i$$

In this expansion the α's will be numerical coefficients fully determined by the nature of ϕ. Now, if the terms of the expanded series occur in a physical context, their summation will be possible only if they are equidimensional irrespective of the value of i. If, then, $\lambda^m \equiv \lambda^n$ for all m and n, it follows that λ, the argument of the function, can only be dimensionless. Our discussion here has been limited to transcendental functions. Algebraic functions such as

$$a = \phi(b, c) = (b + c)^{1/2}$$

present little difficulty. In this case, for instance, we see that the square-root function has derived directly from the manipulation of $a^2 = (b + c)$, all the terms of which will be PPs and equidimensional.

3.3   Dimensional constants

In **1.6** we stated that a constant occurring in a physical equation must either be a pure number, a dimensionless constant or a dimensional constant. It may happen that apparently *ad hoc* dimensions have to be allocated to a constant in order to preserve the homogeneity of an equation. This most often happens in equations of a semi-empirical nature which derive from an incomplete analysis of a problem. And doubts as to how the constant may best be interpreted are usually resolved as soon as a more detailed investigation has been carried out.

Thus in **2.3.1** we expressed the proportionality of density to pressure in a perfect gas as $\rho = k \cdot p$, and noted initially that k appeared as a somewhat mysterious constant having the dimensions of $L^{-2} T^2$. Further investigation, however, showed that k was in fact equal to ρ_0/p_0 and the reason for its particular dimensional structure became evident.

As another example, consider a series of experiments on simply supported beams of different lengths but of the same cross-section. The deflection at the centre under the influence of a weight W is found to be in accordance with the equation $y = k \,.\, Wl^3$, where l is the length of the beam. Now a naive, but dimensionally conscious, physicist would argue that k was clearly a constant with dimensions $M^{-1} L^{-3} T^2$ and he might be puzzled by the appearance of this dimensional structure. An increase in insight, however, shows that although k has indeed these dimensions, it may better be interpreted as a complex quantity derived from the moment of inertia I of the beam cross-section and the Young's modulus E of the material used in the construction. In fact, theoretical analysis shows that $k = 1/(48EI)$ and the reason for its dimensional structure becomes clear.

In the light of these examples, let us consider the gravitational equation: $f = G \,.\, m_1 m_2 / d^2$. The constant G has the dimensions $M^{-1} L^3 T^{-2}$ and it appears at least possible that this may be derived from more fundamental quantities of which we — at least as Newtonian physicists — are ignorant. We are certainly reluctant to regard G as a dimensionless constant of proportionality, although no inconsistency arises if we decide to do so. (See, for example, Burniston Brown[7] and Langhaar[41].) One consequence of regarding G as dimensionless would be that we should have to attribute the dimensions of $L^3 T^{-2}$ to mass. While there is no logical objection to this, we shall find (5.1) that there are excellent heuristic reasons for resisting an approach that involves any decrease in the number of reference dimensions available for analysis.

It is instructive to compare the gravitational equation with the defining equation of force, that is $f = ma$. In this latter case we do not say that force is *proportional* to 'mass times acceleration' but rather that force is *defined and measured* by 'mass times acceleration'; hence the inclusion of a constant in the equation $f = k \,.\, ma$ is no more necessary than would be the inclusion of a constant in $v = ds/dt$, the defining equation of velocity.

We could, had we so wished, certainly have defined f in terms of the gravitational equation by writing $f = m_1 m_2 / d^2$. With this approach the equation $f = k \,.\, ma$ would then be a derived rather than a defining equation and k would be a dimensional constant with the structure $ML^{-3} T^2$. There would be nothing illogical in such a procedure, but it appears artificial and lacks 'taste'. It is a question of developing not only a consistent system, but a system which satisfies certain criteria of simplicity and naturalness. It is evident then that, just as we have a choice in the selection of physical quantities as members of the reference set, so also do we have a choice in the relationships which we use to define derived quantities.

No difficulties are likely to arise provided that the dimensions of all quantities in an equation, other than those of the constant itself, are known. Thus, in the case of Hooke's law defining Young's modulus, $E = \sigma/\epsilon$, we are on firm ground in asserting that the dimensions of E will be those of stress, even though we may admit that the relationship is semi-empirical and has not necessarily been deduced from fundamental considerations concerning the intermolecular forces. But where a defining equation contains both a constant of proportionality and a quantity q, each of undetermined dimensions, then we shall be on doubtful ground if we attempt to obtain a dimensional representation of q by assuming the proportionality constant to be dimensionless.

This remark has applications when, for example, in **8.3** we come to consider the equations defining permeability and permittivity.

3.4 Empirical equations

Where equations are unashamedly empirical, where, that is to say, they are derived from observation unleavened by theoretical principles, then there is no reason to expect homogeneity.

Examples of dimensionally heterogeneous empirical equations are:

$$A = kW^{1/2}/s \tag{1}$$

(Here A is the amplitude of a strain wave in thousandths of an inch caused by the detonation of W pounds of explosive at s feet from the point of observation, k being a constant dependent upon the nature of the rock and soil and varying between 100 and 200.)

$$d = 67.39^h - 0.33 \tag{2}$$

(Here d is the distance in feet at which a road sign will be generally legible, h being the height of the lettering in inches.)

$$E = 10^{11.13}10^{1.8A} \tag{3}$$

(Here E is the energy in ergs released by an earthquake, A being the logarithm of the maximum trace amplitude, expressed in thousandths of a millimetre, and as recorded on a standard seismometer at an epicentral distance of 100 kilometres from the 'quake'.)

All these are dimensionally heterogeneous and it is necessary, therefore, that a definite statement be made as to the units used in the measurement of the various quantities. If units other than those stated be used, then the constants appearing in the equations must clearly be transformed, this being, in contrast to the position of dimensionally homogeneous equations which hold good irrespective of the units used. Note also that the constants appearing in these equations are not the 'neat' numbers characteristic of equations derived from basic principles **(1.6)**.

Equation 3 may profitably be compared with the theoretical relationship:

$$E = 8\pi^3 vtn^2 A^2 \rho d^2$$

where E is the energy released by an earthquake, v is the velocity of the seismic waves, t the time taken by the wave train to pass the observer, n the frequency of the oscillations, ρ the rock density, d the distance from the observer to the 'quake' centre and A the mean amplitude of the seismic wave. This equation, being logically deduced, is homogeneous and, therefore, holds good in whatever system of units may be employed.

Although the position seems tolerably clear, borderlines are inevitably shadowy and it is not always possible to say that *this* equation is properly constructed, *that* equation is empirical. Hooke's law, for instance, is generally regarded as lying within the domain of respectable physics, but it might be difficult to justify this attitude in the face of a sceptical critic. It would not be easy to answer one who argued that the law merely consists of the empirical observation that stress is proportional to strain and that this situation is in no way significantly different to that described, say, by equation 1 which is based upon the equally empirical observation that the strain wave amplitude A is proportional to $W^{1/2}$.

In any case, the homogeneity of any empirical equation may always be restored by the *ad hoc* introduction of dimensional constants. Thus we may impose homogeneity upon even so blatantly empirical an equation as $P = 1000 + 450C$, giving P, the retail price in rupees of a domestic refrigerator in Bombay (1974), where C is the capacity in cubic feet. To restore homogeneity we regard the constant 1000 as having the dimensions of rupees (the limiting price at zero capacity), while the constant 450 would have dimensions equivalent to dP/dC, the price increase per unit increase in capacity.

3.5   Effects of integration and differentiation

The integration of differentiation of an equation will effect the dimensions of the terms, though not the homogeneity of the equation as a whole. Very simply, integration of a function, being equivalent to the determination of the limiting sum of a series of products of that function with δx, results in the dimensions of the function being increased by the dimension of x, while differentiation with respect to x results in the dimensions being similarly reduced.

As an example, consider the differential equation for simple harmonic motion:

$$d^2 x/dt^2 = -\omega^2 x$$

where $\omega \equiv T^{-1}$. Integrating gives

$$x = A \cos(\omega t - \alpha)$$

where the amplitude $A \equiv x$, and we see that integration twice with respect to t has increased the dimensions of both sides of the original equation by a factor of T^2.

The integration of a more complex equation is instructive. Consider the second-order linear equation as applied to a physical situation:

$$A \frac{d^2y}{dx^2} + B \frac{dy}{dx} + Cy = 0 \qquad (1)$$

Each term will be equidimensional and there will be no lack of generality if we consider it as dimensionless. This will imply that $A \equiv x^2/y$, $B \equiv x/y$ and $C \equiv 1/y$. Adopting the standard technique for solving, we form the auxilliary equation

$$Am^2 + Bm + C = 0$$

Here we must have $Am^2 \equiv Bm \equiv C \equiv 1/y$, which implies that $m \equiv 1/x$. Now, if m_1 and m_2 are the two roots, the complete solution of equation 1 will be

$$y = a_1 e^{m_1 x} + a_2 e^{m_2 x}$$

Here a_1 and a_2 will have the dimensions of y and the exponents $m_1 x$ and $m_2 x$ will be dimensionless as required by 3.2.

As a further example, take the partial differential equation

$$\frac{\partial^4 y}{\partial x^4} + \frac{W}{gEI} \cdot \frac{\partial^2 y}{\partial t^2} = 0$$

which represents the nature of the free transverse vibrations set up in a bar of weight W per unit length, x and y being the lengths measured respectively in the axial and transverse directions. (See Rayleigh[56].) A solution of this is

$$y = (A \cos mx + B \sin mx + C \cosh mx + D \sinh mx) . \cos (nt + \alpha)$$

where $m^4 = n^2 W/gEI \equiv L^{-4}$. Here n is the frequency; A, B, C and D are constants of dimension L designed to fit the boundary conditions of the system and α is a pure number. It will be noted that, while the terms of the differential equation are of dimension L^{-3}, the terms of the solution are each of dimension L. The effect effect of integrating this fourth-order equation has been to increase the dimensions of the original terms by a factor of L^4.

Suppose, however, that we rewrite our equation in the form

$$\frac{\partial^2 y}{\partial t^2} + \frac{gEI}{W} \cdot \frac{\partial^4 y}{\partial x^4} = 0$$

Each term is now of dimension LT^{-2} and integration this time results in a dimensional increase of each term by a factor of T^2. The reader may care to effect the (trivial) reconciliation between these results.

Differentiation tends to be even more straightforward than integration. We have, for instance, the equivalence

$$d(ax^n)/dx \equiv n . a . x^{n-1}$$

which follows in view of the fact that n will be a numeric.

We should perhaps mention that there is no anomaly in cases such as $d(\sin x)/dx \equiv \cos x$ because the function $\sin x$ has no physical meaning unless x is dimensionless, in which case the equivalence holds good. If, however, we take x as a dimensional quantity, say $x \equiv L$, then we shall be faced with the differentiation of $\sin ax$ with $a \equiv L^{-1}$. We have then

$$d(\sin ax)/dx = a \cos ax$$

and all is well.

Similarly, partial differentiation provides no problems. As an example, if the characteristic equation for a certain substance is $f(p, v, \theta) = 0$, we have no difficulty in showing that:

$$\frac{\partial p}{\partial v} \cdot \frac{\partial v}{\partial \theta} \cdot \frac{\partial \theta}{\partial p} \equiv [1]$$

3.6 'Mathematical' equations

In addition to 'physical' equations, we find that 'mathematical' equations will be dimensionally homogeneous provided they are susceptible to a geometric interpretation, that is an interpretation in terms of the dimension L rather than in terms of number.

Consider the representation of a parabola:

$$y = p + qx + rx^2$$

The term at the left-hand side consists solely of the ordinate y, of dimension L. It seems reasonable to expect that the remaining terms will be dimensionally equivalent and this is indeed the case for:

1. p has the dimension L, being the length of the intercept of the parabola on the y-axis when $x = 0$.

2. q represents the gradient, dy/dx, at the point where $x = 0$. This has the dimension L/L and therefore the term qx has the dimension L.

3. r is inversely proportional to the radius of curvature, $1/2r$, of the curve at $x = 0$ and has therefore the dimension L^{-1}, implying that the term rx^2 has, once again, the dimension L.

Dimensional homogeneity is unaffected by manipulation in 'mathematical' as well as in physical equations unless, of course, non-homogeneous quantities are added to each side. Consider the quadratic equation $ax^2 + bx + c = 0$. Regarding x as having the dimensions of L, we induce homogeneity by putting $a \equiv L^{-1}$, $b \equiv L^0 \equiv [1]$ and $c \equiv L$. We may then confirm that in the solution to the equation homogeneity will be maintained, for we have

$$x = \frac{-b + \sqrt{(b^2 - 4ac)}}{2a} \equiv L$$

Testing the homogeneity of a few integrals, selected at random from a table, affords an interesting exercise. We may readily see by inspection that

$$\int x^m \cdot dx \equiv x^{m+1}/(m+1) \quad \text{(for homogeneity } m \equiv [1])$$

$$\int \frac{dx}{\sqrt{(a^2 - x^2)}} \equiv \arcsin (x/a) \quad \text{(for homogeneity } a \equiv x)$$

$$\int \frac{dx}{a + bx} \equiv \frac{1}{b} \log (a + bx) \quad \text{(for homogeneity } a \equiv [1], b \equiv x^{-1})$$

$$\int x e^{ax} \cdot dx \equiv \frac{e^{ax}}{a^2} (ax - 1) \quad \text{(for homogeneity } a \equiv x^{-1})$$

This persistence of dimensional homogeneity may be useful in checking a result. Thus, in the last example, had the right-hand side been inadvertently written as, say,

$$\frac{e^{ax}}{a}(ax - 1)$$

then the error would have been immediately apparent.

The requirement of a 'geometrical' interpretation is essential. We see, for instance, that the binomial expansion

$$(a + x)^n = a^n + n \cdot a^{n-1}x + \tfrac{1}{2}n \cdot (n - 1)a^{n-2}x^2 + \ldots$$

is homogeneous provided that $a \equiv x \equiv L$ and, in this case, the expression $(a + x)^n$ represents the volume of a hypercube in n space of side length equal to $(a + x)$. But, in contrast to this, we see that the expansion

$$(1 - x)^{-1} = 1 + x + x^2 + \ldots$$

is homogeneous only if x is considered as a numeric or, alternatively, if the term 1 occurring on the left-hand side is regarded as a unit length. In this latter case, to maintain homogeneity, the suppressed coefficient of each term of the series must be made explicit, and we write

$$(1 - x)^{-1} = 1^{-1} + 1^{-2} \cdot x + 1^{-3} \cdot x^2 + \ldots$$

Where, as in the equations of number theory or analysis, no such dimensional attribution is made, no homogeneity need occur. There are, nevertheless, a number of intriguing aspects of homogeneity as applied to purely 'mathematical' equations, but we refrain from pursuing these since our book is primarily designed to deal with dimensional problems in engineering and physics only.

We do, however, make one final point. In a vector equation, each term and subterm — that is, each group to be added or equated — will be either a vector or a scalar. Thus the equation for a plane passing through the point r_0 and containing the vectors s and t, is:

$$(r - r_0) \cdot (s \times t) = 0$$

Note that in the left-hand bracket, the two subterms are vectors and the difference, therefore, is a vector. The product in the right-hand bracket is a vector. The left-hand term as a whole, then, being a scalar product will be scalar. It follows that the zero on the right-hand side must be interpreted as a number rather than as a zero vector.

4

Systematic Calculation of Dimensionless Products

4.1 First method – determination of DPs by inspection: wavelength of gravity waves

It will be recalled from 2.4 that a dimensionally homogeneous physical equation may be written in the form $f(\pi_1, \pi_2, \ldots, \pi_k) = 0$, where the π's represent a complete set of DP's constructed from the relevant quantities entering into a given situation. Moreover k, the number of DPs present, will generally be equal to $(n - r)$, where n is the number of variables and r the number of reference quantities or dimensions.

Three methods are conveniently available for the systematic calculation of sets of DPs. These methods will be given in an order which affords increasing depth of insight into the process involved and they will be demonstrated by the use of examples rather than formally.

In many cases the set of DPs may be determined by inspection. Suppose we are interested in determining the equation descriptive of the wavelength of deepwater 'gravity' waves. Our understanding of the situation suggests that the only significant physical variables will be the wavelength itself, the wave velocity, the gravitational acceleration and, possibly, the density of the water. We accordingly construct the following table listing these quantities and showing the exponents of M, L and T occurring in their dimensional formulae:

Physical quantity	Symbol	M	L	T
Wavelength	λ	0	1	0
Wave velocity	v	0	1	-1
Gravitational acceleration	g	0	1	-2
Water density	ρ	1	-3	0

From an inspection of this table we see that:

1. If a product π_1 is to be dimensionless in M it cannot contain the quantity ρ.

2. If π_1 is to be dimensionless in T, then v and g must be associated in the product

in some such form as $g/v^2 \equiv (L/T^2) \cdot (T^2/L^2)$, for only then will the exponent of T become zero.

3. If, now, we take π_1 as containing v and g in the form g/v^2, then π_1 can be rendered dimensionless in L only provided that λ enters as a multiplier of g. π_1 will, in consequence, take the form

$$g\lambda/v^2 \equiv (L/T^2) \cdot L \cdot (T^2/L^2) \equiv [1]$$

Since there are 4 physical quantities which are subject to 3 restrictions imposed upon them by the necessity of making the dimensions of M, L and T each equal to zero, there will, according to Buckingham's theorem, be $(4 - 3) = 1$ DP only in a complete set. We may, then, take this DP as

$$\pi_1 = (g\lambda/v^2)$$

and the required relationship between the quantities will be simply

$$f(g\lambda/v^2) = 0$$

It follows, since f is constantly equal to zero, that the argument of f must also be constant, that is $g\lambda/v^2 =$ constant, or

$$\lambda = k \cdot v^2/g$$

giving the wavelength in terms of the other variables as required. This represents a complete solution to our problem since there occurs in the final equation no undetermined function, but only the undetermined coefficient k.

A few comments. The numerical value of the constant k cannot in principle be determined if the argument is to be restricted to dimensional considerations (11.3). If, however, we had carried out a full analysis by conventional methods, we should have found that $\lambda = 2\pi v^2/g$ and, in consequence, $k = 2\pi$.

Had we argued that v and g could occur in association with one another in some alternative form, say v^2/g (rather than g/v^2), our functional equation would have become $f(v^2/g\lambda) = 0$. This, or indeed $f((v^2/g\lambda)^n) = 0$ for any positive or negative value of n, would clearly have yielded the same solution as was, in fact, obtained.

We are free to make an arbitrary choice between 'equivalent' quantities in the original selection of variables. Thus, if we had chosen the wave frequency n instead of the wavelength λ, the analysis would have proceeded along almost identical lines and, since $\lambda = v/n$, we should have obtained as our final relationship $n = k \cdot g/v$, which is equivalent to the result as originally deduced.

Similarly, in a problem involving, say, the behaviour of a falling raindrop, it would be of no consequent whether we chose the radius r of the drop, the surface area A, or the volume v as the significant variable to be considered. If the relationship were such than an equation containing r^a were involved, then, had we conducted our argument in terms of A or v, the final equation would have contained $A^{a/2}$ or $v^{a/3}$ respectively instead of r^a.

We next reconsider our example with the additional assumption that the water is sufficiently shallow for the depth d significantly to effect wave behaviour. There will now be 5 variables subject to the same 3 restrictions, giving $(5 - 3) = 2$ DPs. We still have $\pi_1 = (g\lambda/v^2)$ as one of the products. In order to obtain the second we argue: if π_1 and π_2 are to form a complete set, π_2 must be independent of π_1 and must, therefore, contain the new variable d of dimension L. The most obvious possibility for such a product is simply: $\pi_2 = (d/\lambda) \equiv L/L \equiv [1]$ and the functional equation will then become

$$f(\pi_1, \pi_2) = f(g\lambda/v^2, d/\lambda) = 0 \tag{1}$$

Solving this for $(g\lambda/v^2)$ yields

$$g\lambda/v^2 = \phi(d/\lambda)$$

or $\quad \lambda = (v^2/g)\, \phi\,(d/\lambda)$ $\hfill (2)$

This solution well illustrates typical difficulties which are liable to arise in dimensional analysis. We note, for instance, as in equation 1 of **2.5**, that the solution is not complete since it contains an undetermined function ϕ which, in fact, may be shown to be

$$\phi\,(d/\lambda) = \frac{2\pi}{\tanh 2\pi(d/\lambda)}$$

But this last result is only to be obtained by recourse to conventional analysis.

Note that we have failed to obtain an explicit relationship for λ, since we chose our set of DPs in such a way that each of them contained the dependent variable. Suppose, now, that we choose for π_2 some other DP containing d, say

$$\pi^*_2 = (gd/v^2)$$

Our functional equation will in consequence take the form

$$f^*\,(\pi^*_1, \pi^*_2) = f^*(g\lambda/v^2, gd/v^2) = 0 \tag{3}$$

which may be solved explicitly for λ to give

$$\lambda = \frac{v^2}{g}\phi\!\left(\frac{gd}{v^2}\right)$$

Once again, however, this is not too useful a result, for all the independent variables now enter the argument of the undetermined function.

We point out that since both (π_1, π_2) and (π^*_1, π^*_2) are complete sets it should be possible to express equation 3 in terms of equation 1. This is indeed the case, for we find that

$$f^* \left(\frac{g\lambda}{v^2}, \frac{gd}{v^2} \left(\frac{g\lambda}{v^2} \right)^{-1} \right) = 0 = f \left(\frac{g\lambda}{v^2}, \frac{d}{\lambda} \right)$$

Although we may not have obtained a complete solution to the second part of our problem, we have nevertheless imposed rigid restrictions upon it, and positive information has been obtained. Thus we see that for conditions in which d becomes a significant variable:

1. λ remains independent of ρ,

2. Consideration of equation 2 shows that if d and λ are both increased by a factor k, the result leaves ϕ and, in consequence. $g\lambda/v^2$ unchanged. It follows that v will then be increased by a factor of $k^{1/2}$.

In view of the practical convenience of this method of calculation, a further and more complex example involving determination of DPs by inspection is given in **5.3.3**.

4.2 Second method – solution of indicial equations: wavelength of gravity waves (*continued*)

A second and more systematic method of calculation is as follows. In order to facilitate comparison with the first method, we continue to use the example of **4.1** and we denote the powers to which the quantities λ, v, g, ρ and d must be raised in a DP respectively by x_1, x_2, x_3, x_4 and x_5. Then in any product of these quantities the conditions to be satisfied if that product is to be dimensionless in M, L and T are simply that the sum of the entries in each of the three columns of the array set out below will be equated to zero.

	M	L	T
λ		x_1	
v		x_2	$-x_2$
g		x_3	$-2x_3$
ρ	x_4	$-3x_4$	
d		x_5	

This array may be written down immediately from an inspection of the table showing the dimensions of the concerned quantities set out in **4.1** to which the further quantity d has now been added. The 'indicial' equations resulting from the requirement that in any DP the exponents of M, L and T will each be zero are then:

$$
\begin{array}{c|ccccc}
 & \lambda & v & g & \rho & d \\
\hline
M & & & & x_4 & = 0 \\
L & x_1 & + x_2 & + x_3 & -3x_4 & + x_5 & = 0 \\
T & & -x_2 & - 2x_3 & & & = 0
\end{array} \qquad (1)
$$

These three equations contain 5 unknowns; we may therefore expect, in general, to be able to solve for 3 unknowns, say x_3, x_4 and x_5, expressing them in terms of the 2 remaining unknowns x_1 and x_2. A little algebraic manipulation shows that

$$
\left.
\begin{array}{l}
x_3 = \qquad -\tfrac{1}{2}x_2 \\
x_4 = 0 \\
x_5 = -x_1 -\tfrac{1}{2}x_2
\end{array}
\right\} \qquad (2)
$$

This leaves us free to select any two sets of arbitrary values for x_1 and x_2, secure in the knowledge that we shall clearly obtain DPs as soon as the remaining indices x_3, x_4 and x_5 have been determined by the substitution in equation 2 of whatever values we may select.

We take as our two sets of values:

$$
\left.
\begin{array}{ll}
1. & x_1 = 1 \quad : \quad x_2 = -2 \\
2. & x_1 = -1 \quad : \quad x_2 = 0
\end{array}
\right\} \qquad (3)
$$

Substitution now leads respectively to the two solutions indicated in the array

	x_1	x_2	x_3	x_4	x_5
π_1	1	-2	1	0	0
π_2	-1	0	0	0	0

Raising λ, v, ... to the powers corresponding to the values we have obtained for x_1, x_2, ... now gives the set of two DPs for which we are seeking, namely

$$\pi_1 = (g\lambda/v^2) \quad \text{and} \quad \pi_2 = (d/\lambda)$$

which are identical with those obtained in **4.1**.

Our choice of values for x_1 and x_2 was admittedly made in such a way as to produce the same forms for π_1 and π_2 as had previously been obtained. We are, however, free to select any values we wish for substitution in equation 2 and this freedom of choice corresponds precisely with the freedom in the construction of DPs which has already been discussed. As before, the final result will be unaffected by the nature of the values selected.

In illustration of this, suppose that, in place of the two sets of values given in equation 3, we choose:

1. $x_1 = 1 : x_2 = 0$

2. $x_1 = 0 : x_2 = 1$

We then obtain for our set of DPs: π^*_1 (λ/d) and $\pi^*_2 = (v/(gd)^{1/2})$ which are, of course, expressible as functions of the DPs previously determined since (π_1, π_2) as well as (π^*_1, π^*_2) represent complete sets. In fact we notice that

$$\pi^*_1 = \pi_2^{-1}$$
$$\pi^*_2 = (\pi_1 \cdot \pi_2)^{-1/2}$$

This result is consistent with the requirement that the variables must be related by an equation of the form $f(\pi_1, \pi_2) = 0$ for, by arguing again as in **4.1**, we see that this equation includes as a special case

$$f(\pi_2^{-1}, (\pi_1 \cdot \pi_2)^{-1/2}) = 0$$

which has just been shown to be equivalent to $f(\pi^*_1, \pi^*_2) = 0$. Similarly $f(\pi_1, \pi_2) = 0$ accommodates all other results containing DPs which may be obtained from the substitution in equation 1 of any other arbitrary values for x_1 and x_2.

There is, however, one proviso. The values allotted to x_1 and x_2 must be independent, or the resulting set of DPs will not be complete. Suppose that one set of arbitrary values is a linear combination of the other set and that we have, for example,

1. $x_1 = 1$: $x_2 = 2$

2. $x_1 = k$: $x_2 = 2k$

where k is any constant. If we now follow the procedure already developed we shall find that

$$\pi_1 = \frac{v\lambda^2}{(gd)^{5/2}} \quad \text{and} \quad \pi_2 = \left(\frac{v\lambda^2}{(gd)^{5/2}}\right)^k$$

which clearly do not represent a complete set.

4.3   Third method – the indicial matrix: energy of a vibrating wire

To illustrate the third method of approach we consider a fresh example dealing this time with the energy of a vibrating wire. Here the table of relevant physical quantities is:

Physical quantity	Symbol	M	L	T
Energy of wire	E	1	2	−2
Length of wire	l	0	1	0
Linear density of wire	ρ	1	−1	0
Amplitude of antinode	A	0	1	0
Tension in wire	f	1	1	−2

Note that in constructing this table we have omitted the frequency of vibration, since this will itself be a function of l, ρ and f and it need not, therefore, be explicitly included.

We first turn our attention to the 'indicial matrix' in which each entry corresponds to that given in the table of quantities:

$$
\begin{array}{c|ccc}
 & M & L & T \\
\hline
E & 1 & 2 & -2 \\
l & 0 & 1 & 0 \\
\rho & 1 & -1 & 0 \\
A & 0 & 1 & 0 \\
f & 1 & 1 & -2
\end{array}
$$

Recall now that the rank of a matrix is defined as the order of the largest non-zero determinant contained therein. In this case we note that the value of the upper third-order determinant in the matrix is non-zero and the rank of our matrix is therefore 3. With this in mind we make use of the well-known result from the theory of linear equations (see, for example, Ferrar [24]) which states that where there are n variables in a set of homogeneous linear equations and where the rank of the corresponding matrix is r, there will then exist a complete* set of $(n - r)$ linearly independent solutions to the equations.

The equations we have in mind here are the indicial equations, formed as in equation 1 of 4.2, by regarding each entry in the matrix as the coefficient of x_1, $x_2, x_3, \ldots$, etc. and by equating the sum of each column to zero, thus ensuring that any power product formed from the n quantities and based on the solution will be dimensionless in M, L and T and, therefore, a DP.

* Ferrar uses the word 'fundamental' rather than 'complete'.

In our present example, the set of homogeneous linear 'indicial' equations will then be

	E	l	ρ	A	f	
M	x_1		$+x_3$		$+x_5$	$= 0$
L	$2x_1$	$+x_2$	$-x_3$	$+x_4$	$+x_5$	$= 0$
T	$-2x_1$				$-2x_5$	$= 0$

$$(1)$$

The result just quoted implies that there will be a complete set of $(n - r) = (5 - 3)$ = 2 linearly independent solutions and, therefore, a complete set of 2 DPs which may be based on these solutions. Note that r now represents the *rank* of the indicial matrix and not necessarily the number of reference dimensions.

Before proceeding we emphasise the importance of this result, for it is basic to much of what is to follow. Indeed we shall find that quite frequently there arises an unsuspected linear dependence between the columns of the indicial matrix and, therefore, between the exponents of the reference dimensions. In such situations a singularity develops and, in consequence, the reference dimensions chosen will not constitute a complete set.

This will be taken up in detail at a later stage, particularly in Chapter 5, and a few simple illustrations of singularities will also be given towards the end of the present section. Meanwhile we say that where no singularity develops in the indicial matrix as a result of linear relationships between the exponents of the reference dimensions chosen, we may refer to that set of dimensions as an 'essential' set for the situation considered. This definition is strictly redundant, in that no linear dependence will exist if a reference set of dimensions is 'complete' in the sense of 1.7. It will, nevertheless, be useful to talk of an 'essential' set when we wish to place a special emphasis upon the first criterion of completeness, that is upon the aspect of completeness which involves independence.

With this behind us we return to the indicial equations 1 which have the solution

$$
\left.
\begin{aligned}
x_3 &= 0 \\
x_4 &= -x_1 -x_2 \\
x_5 &= -x_1
\end{aligned}
\right\}
\tag{2}
$$

We are, as in **4.2**, now free to choose two sets of arbitrary values for x_1 and x_2 and we select

$$
\left.
\begin{aligned}
1. \quad & x_1 = 1 \ : \ x_2 = 0 \\
2. \quad & x_1 = 0 \ : \ x_2 = 1
\end{aligned}
\right\}
\tag{3}
$$

Substitution in equation 2 shows that the corresponding values of the remaining indices will be as displayed in the 'matrix of solutions':

	x_1	x_2	x_3	x_4	x_5
π_1	1	0	0	−1	−1
π_2	0	1	0	−1	0

The set of two DPs based upon these two independent solutions will then be

$$\pi_1 = (E/Af) \quad \text{and} \quad \pi_2 = (l/A)$$

and the procedure previously adopted yields the final equation

$$E = Af\,\phi\,(l/A)$$

which shows that for fixed length and amplitude, the energy in a vibrating wire varies directly with the tension.

It may, at first sight, seem surprising that the density of the wire has dropped out. This apparent anomaly arises from the fact that although for fixed *frequency* E varies directly with the density, for fixed *tension* the frequency itself varies inversely with the density. With the set of variables we have selected, these two effects are compensatory and E is not, therefore, a function of density.

The method of the present section is effectively that of **4.2** with the added refinement provided by the investigation of the rank of the indicial matrix. Where this investigation shows that the matrix is singular, that its rank is less than the number of columns corresponding to the reference dimensions, this will make no difference in principle to the technique of calculation. The number of DPs that are thrown up will remain $(n - r)$ and the effect of the singularity will merely be to increase the number of DPs necessary to form a complete set.

Indeed, a singular matrix makes its appearance as soon as we attempt to work with more reference dimensions than the situation warrants − as when we introduce the dimension M into a purely kinematical problem. As an illustration, suppose that we wish to determine the distance fallen by a body from rest under the influence of gravity. The indicial matrix would then be

	M	L	T
s	0	1	0
t	0	0	1
g	0	1	−2

which is, of course, singular, being of rank 2. (In a trivial sense, the first column is linearly dependent upon the other two; the exponent of M is always zero times that of L plus zero times that of T.) In consequence there will be $(3 - 2) = 1$ DP which turns out to be $\pi_1 = (s/gt^2)$. This is substituted in the equation $f(\pi_1) = 0$ and we find as our solution (s/gt^2) = constant, or $s = k \cdot gt^2$.

Note that equality between n and r is possible. In such cases $(n - r) = 0$ and there will be no significant solution since the rows of the matrix will then be linearly independent and no one quantity can be expressed in terms of the others. Consider, for instance, the matrix based on the three quantities weight, density and velocity:

$$
\begin{array}{c|ccc}
 & M & L & T \\
\hline
W & 1 & 1 & -2 \\
\rho & 1 & -3 & 0 \\
v & 0 & 1 & -1 \\
\end{array}
$$

The rank is 3 and, since $(n - r) = 0$, it follows that no DP can be formed and, in consequence, no relationship between the variables is possible.

Contrast this with the situation shown in the matrix based on the three quantities force, mass and acceleration:

$$
\begin{array}{c|ccc}
 & M & L & T \\
\hline
f & 1 & 1 & -2 \\
m & 1 & 0 & 0 \\
a & 0 & 1 & -2 \\
\end{array}
$$

This is singular, the rank is 2 and, since $(n - r) = 1$, there exists 1 DP, $\pi_1 = (f/ma)$, which corresponds to the relationship $f = ma$. Note that the singularity here results from the linear dependence of the L and T columns. L occurs only in conjunction with T^{-2} and, had we wished, we could have worked with the single 'composite' dimension LT^{-2}. Or, with alternative phrasing, we may say that, in this particular situation, an essential set of dimensions could be based upon two quantities only: mass and the 'derived' quantity acceleration.

This example shows that a reference set may be essential (or complete) in one situation but not necessarily so in another, as follows from the fact that it is always possible to make a restricted selection of variables in such a way that dependence is imposed upon dimensions that are generally independent of one another. It is, indeed, often difficult to determine whether or not this situation arises at the start of a problem, and it is for this reason that the approach involving an examination of the rank of the indicial matrix is so essential to the rigorous approach to dimensional analysis.

This last point is fully consistent with Buckingham's pi-theorem, provided that we insist, as we did in our formulation of the theorem, that r is the number of *independent* (essential) reference dimensions. Alternatively, we may rephrase the theorem simply by defining r as the rank of the indicial matrix, rather than as the number of members in the reference set.

The occurrence of a singularity in the indicial matrix may sometimes be interpreted by observing that the solution to the problem considered is simply not expressible as a single PP relationship (8.5.5). If, for instance, the solution involves terms of the form $a = b + c$, then this entails the existence of two DPs and the rank of the matrix will be reduced to accommodate the situation. Analysis would then give $f(\pi_1, \pi_2) = 0$, with $\pi_1 = (b/a)$ and $\pi_2 = (c/a)$. It is to be expected, then, that if the number of variables exceeds the number of reference dimensions by only one, a linear relationship between the columns of the matrix must arise for, failing this, the required singularity will not occur.

It is, of course, never possible for the value of $(n - r)$ to be less than zero, since the rank of a matrix cannot exceed the number of rows.

We have carried out this discussion almost entirely in terms of the columns of the matrix. There is, however, a well-known result which tells us that if there is a linear independence between the columns of a determinant, then so also will there be one between the rows. (See, for example, Littlewood [43].) To clarify the present application of this, we suppose that we are faced with an example in which the indicial matrix is singular and of rank r. This implies that any determinant contained within that matrix and of order $(r + 1)$ will have the value of zero and the corresponding $(r + 1)$ rows, therefore, will be linearly dependent. It follows that under such conditions every set of $(r + 1)$ variables will be linearly related. If, then, the total number of variables $n > (r + 1)$, we lose nothing by taking advantage of such relationships to make an appropriate reduction in the number of variables considered and, as we shall see in **9.1**, this approach will enable us to add precision to the result of an analysis.

As a point of practical importance, we return to the selection of arbitrary values for substitution in the indicial equations. Suppose $(n - r) = 3$ and that we have to select three sets of arbitrary values for the indices x_1, x_2 and x_3. It will, then, in general, be convenient to put:

1. $x_1 = 1$: $x_2 = 0$: $x_3 = 0$

2. $x_1 = 0$: $x_2 = 1$: $x_3 = 0$

3. $x_1 = 0$: $x_2 = 0$: $x_3 = 1$

This procedure ensures that the three variables corresponding to x_1, x_2 and x_3 each enter into one and only one of the three DPs of the complete set. It will often be useful to isolate in this fashion the three variables in which we happen to be most interested and, by adopting the recommended technique, it may be possible to determine explicitly the manner of dependence of one or other of these variables upon the remaining quantities. This has been phrased to meet the case where $(n - r) = 3$. The argument is, however, general.

Thus, in the first example of this section, involving the energy of a vibrating wire, we had $(n - r) = 2$ and we chose as our two sets of arbitrary values $(x_1 = 1 : x_2 = 0)$ and $(x_1 = 0 : x_2 = 1)$, thus arriving at a useful result. Had we, on the other hand put, say, $(x_1 = 1 : x_2 = 0)$ and $(x_1 = x_2 = 1)$, the resultant DPs would have been

$$\pi_1 = (E/Af) \text{ and } \pi_2 = (El/A^2 f)$$

We should then have been unable directly to determine the dependence of E upon f, since both these variables would have entered into the argument of the undetermined function. As a result we would have been faced with the equation $E = Af \cdot \phi(El/A^2 f)$, which is largely uninformative.

Alternatively, it will be possible to manipulate the DPs in order to increase the information they yield. Thus if, as in the preceding paragraph, we had obtained $\pi_1 = (E/Af)$ and $\pi_2 = (El/A^2 f)$, it would have been profitable to notice that since π_1 and π_2 comprise a complete set of DPs, so also do π_1 and (π_1/π_2), the latter set reducing to the two DPs originally obtained, that is to (E/Af) and (l/A).

This point may be re-expressed in terms of dependent and independent variables. What we aim to accomplish is the isolation in one DP only of the dependent variable and such independent variables as we may be most interested in, leaving the independent variables of lesser interest to comprise such other DPs as may be needed to complete the set.

4.4   Third method – a further example: thrust developed by a ship's propeller

In view of the importance of the method involving the use of the indicial matrix, we set out one further and more complex example of its use. This example, taken from Buckingham[8], considers the thrust f of a ship's screw propeller which we consider as dependent upon the density of the water ρ, the radius of the screw r, the linear velocity of the ship v, the revolutions per unit time n, the viscosity of the water μ, and the acceleration due to gravity g.

The indicial matrix is

	M	L	T
f	1	1	-2
ρ	1	-3	0
r	0	1	0
v	0	1	-1
n	0	0	-1
μ	1	-1	-1
g	0	1	-2

(We now abandon the more usual method of writing the matrix enclosed by parallel vertical lines; the form here introduced is considered more convenient for our special purposes and will be used in the remainder of the book.)

The matrix is of rank 3, with $n = 7$ variables, we expect to find $(7 - 3) = 4$ DPs in a complete set, these being based on the 4 linearly independent solutions to the indicial equations. These equations are

	f	ρ	r	v	n	μ	g	
M	x_1	$+x_2$				$+x_6$		$= 0$
L	x_1	$-3x_2$	$+x_3$	$+x_4$		$-x_6$	$+x_7$	$= 0$
T	$-2x_1$			$-x_4$	$-x_5$	$-x_6$	$-2x_7$	$= 0$

We solve in terms of x_1, x_2, x_3 and x_4, obtaining

$$x_5 = 3x_1 - 3x_2 + 2x_3 + x_4$$

$$x_6 = -x_1 - x_2$$

$$x_7 = -2x_1 + 2x_2 - x_3 - x_4$$

Now take the arbitrary values of x_1, x_2, x_3 and x_4 as

1. $x_1 = 1 : x_2 = 0 : x_3 = 0 : x_4 = 0$
2. $x_1 = 0 : x_2 = 1 : x_3 = 0 : x_4 = 0$
3. $x_1 = 0 : x_2 = 0 : x_3 = 1 : x_4 = 0$ $\qquad$ (1)
4. $x_1 = 0 : x_2 = 0 : x_3 = 0 : x_4 = 1$

and from these values the matrix of solutions may be readily prepared:

	x_1	x_2	x_3	x_4	x_5	x_6	x_7
π_1	1	0	0	0	3	-1	-2
π_2	0	1	0	0	-3	-1	2
π_3	0	0	1	0	2	0	-1
π_4	0	0	0	1	1	0	-1

The functional equation which we wish to determine is then

$$f(\pi_1, \pi_2, \pi_3, \pi_4) = f\left[\frac{fn^3}{\mu g^2}, \frac{\rho g^2}{n^3 \mu}, \frac{rn^2}{g}, \frac{vn}{g}\right] = 0$$

Note that following our schema of values for x_1, x_2, x_3 and x_4, the variables f, ρ, r and v each appear once only and are, moreover, confined to separate DPs. As a final step we express our result explicitly for f in terms of the remaining variables:

$$f = \frac{\mu g^2}{n^3} \cdot \phi \left[\frac{\rho g^2}{n^3}, \frac{r n^2}{g}, \frac{v n}{g} \right]$$

We have now attained a position in which we can show almost intuitively that $(n - r)$, the number of DPs obtained, represents a complete set. We originally started with n 'unknowns' representing the indices of our variables and connected by r indicial equations. Each of these equations may be used in turn to eliminate one of the unknowns, leaving us with a final solution in terms of $(n - r) = 4$ 'unknown' indices only, and we see that we have expressed each of x_5, x_6 and x_7 in terms of x_1, x_2, x_3 and x_4. We then proceeded to allocate 4 sets of arbitrary values to these latter indices as shown in schema 1, adopting the procedure of ensuring that in each set of values one index is put at unity with the remainder being put at zero.

As a result it is simple to see that:

1. Were we to have allocated s sets of values with $s < 4$, then, in general, there would exist other sets which could not be linearly derived from the s sets since at least one of the indices would only occur with zero value.

2. Were we to have allocated t sets of such values with $t > 4$, then each such set in excess of 4 could be readily expressed as a linear combination of the 4 sets only.

Taking into consideration the manner in which the DPs are based upon the values allocated to x_1, x_2, x_3 and x_4, we see that a complete set of DPs will involve the allocation of $(n - r)$ sets of values to the $(n - r)$ unknowns which remain after the solution of the indicial equations.

This has been argued for the case where $(n - r) = 4$, but the argument is of general application, as may be seen by applying it to the example of **4.3** where $(n - r) = 2$ and where, in consequence, the number of members of a complete set of allocations to the unknowns x_1 and x_2 was 2, as was the number of DPs which were based on that set. (See schema 3 of **4.3**.)

This does not represent a formal proof that the number of members in a complete set of DPs is $(n - r)$. Our intention has been merely to set out an argument in such a manner as to provide the reader with a basic understanding of the principles underlying the situation. The rigorous proof must be based on the theory of linear equations, a reference to which has already been provided.

4.5   Some miscellaneous comments

We conclude with a number of miscellaneous comments. Firstly, we should never say that dimensional analysis has in some way 'failed' if it yields something short of a complete solution involving only an undetermined constant and no undetermined function. If, starting with n variables, we end with a relationship between $(n - r)$ DPs, there has been a real gain in information, for the DPs may be regarded as new variables in their own right and the fact that we have been able to effect a reduction in the number of those originally appearing is of real importance.

If, for example, the result of an analysis involves 2 DPs with $f(\pi_1, \pi_2) = 0$, then we may obtain full information concerning the situation by plotting one DP against the other (see figure 18, page 152). Where we are left with 3 DPs we may fully summarise the situation by a single contour diagram (see figure 10, page 102), and even with 4 DPs, as in the example of **4.4**, we are still left with a considerably more manageable result than we were faced with initially. A further point, which we consider in Chapter 10, is that the experimental investigation of a problem becomes greatly simplified if it can be based on a small number of DPs rather than upon a relatively large set of independent variables.

Note next that a single set of variables may be related to a variety of physical situations. In such cases the same DP or DPs may have a number of differing interpretations. Suppose we consider that the quantities t, g and l are functionally dependent upon each other, we shall then obtain 1 DP only, that is $(t^2 g/l)$, yielding the equation

$$t = k\sqrt{(l/g)}$$

This equation, with $k = 2\pi$, gives the period of a simple pendulum; with $k = \sqrt{2}$ it gives the time of free gravitational fall of a body over a distance l; and with $k = \sqrt{(2\pi)}$ it gives the period of a deepwater wave of wavelength l. With other appropriate values of k, it would necessarily apply to any further situation in which the quantities t, l and g, and only those quantities, were related.

Dimensional analysis, if unskillfully used, may result in the introduction of complexities rather than in their elimination. Consider the relationship between the acceleration of a body, its initial and terminal velocities, its distance and its time of travel. We have here $n = 5$ variables and, since M does not enter into the situation, the rank of the matrix will be $r = 2$ giving $(n - r) = 3$ DPs. With the usual notation, the functional equation turns out to be

$$f\left(\frac{s}{vt}, \frac{as}{v^2}, \frac{u}{v}\right) = 0$$

or, since we are regarding a as the dependent variable,

$$a = \frac{v^2}{s} \phi\left(\frac{s}{vt}, \frac{u}{v}\right)$$

It will not be immediately obvious that the function ϕ is $\phi = (s/vt)(1 - u/v)$, this

being entailed by the solution which, in the familiar form, we recall as $a = (v - u)/t$.
Finally we return to the question of scale changes briefly mentioned in 1.6. To
fix our ideas, let us consider the period t required for a small oscillation of a
Nicholson hydrometer of mass m and neck cross-section A floating in a liquid of
density ρ. The indicial matrix is seen to be

	M	L	T
t	0	0	1
m	1	0	0
A	0	2	0
ρ	1	-3	0
g	0	1	-2

Without now concerning ourselves with the solution to the problem, which may be
left as a simple exercise, we suppose that the variables, measured in one system of
units have the magnitudes t, m, A, ρ, and g, while in some other system of units
they have the magnitudes t', m', A', ρ', and g'. Then the equations of transformation between the magnitudes of the sets of variables measured in the two
systems will be simply

$$
\left.
\begin{aligned}
t' &= (M^0 L^0 T^1) \cdot t \\
m' &= (M^1 L^0 T^0) \cdot m \\
A' &= (M^0 L^2 T^0) \cdot A \\
\rho' &= (M^1 L^{-3} T^0) \cdot \rho \\
g' &= (M^0 L^1 T^{-2}) \cdot g
\end{aligned}
\right\} \quad (1)
$$

Here the exponents of the members of the reference set correspond to the entries in
the indicial matrix already set out. If, now, $t, m, \ldots$ be measured in, say, SI units,
while $t', m', \ldots$ be measured in foot–pound–second units, we should use as our
'conversion factors' for substitution in equations 1 the values

$M = 2.205$ (kg/lb)
$L = 3.281$ (m/ft)
$T = 1.000$

● The set of equations 1 can readily be shown to represent a three-parameter
transformation group. In confirmation of this, note that:

1. The identity transformation exists with $M = L = T = 1$.

2. The inverse transformation, taking $t', m', \ldots$ back to $t, m, \ldots$ exists.

3. If in any third system of units the transformation of $t, m, \ldots$ to $t'', m'', \ldots$ be

made, then the transformation from $t\,'', m''', \ldots$ to $t', m', \ldots$ will be of the same form as equations 1 and the requirement of closure will be satisfied.

Some recent authors have found it profitable to develop group theoretic aspects of such sets of equations as equations 1 in some depth. While we do not ourselves propose to follow this approach, the interested reader may, for example, be referred to papers by Hainzl[31], by Moran[50] and by Na and Hansen[51]. ●

5

Extensions to the Set of Reference Dimensions

5.1 Introduction

Knowledge concerning the interrelationship between physical quantities entering into a situation maximises when the number of DPs minimises — preferably to one, in which case we obtain a complete solution. Since the number of DPs in a set is $(n - r)$, and since n is usually fixed, we may most readily add to the information available by increasing r, the rank of the indicial matrix. Now it generally happens that r increases with the number of columns, that is with the number of reference dimensions, and we are, in consequence, led to consider the question of determining the maximum number of dimensions which may be used in a given case.

It is not profitable to ask what is the 'true' number of reference dimensions. We have already seen that we could, if we so wished, define without logical inconsistency all quantities in terms of L and T (3.3). We could, indeed, write all equations in terms of only one dimension by, for example, putting $M \equiv L \equiv T \equiv X$, and the property of dimensional homogeneity would still persist in physical equations. But this approach would be sterile and unheuristic. We prefer, therefore, to enquire what is the *maximum* number of dimension which may be operated on — again without logical inconsistency.

An attempt to increase the number of dimensions in the reference set is seen in problems where electrical or thermal quantities are encountered and where the basic set MLT is extended to include, say, Q ($\equiv$ electric charge) or Θ ($\equiv$ temperature). Again, under certain circumstances MLT may be extended to embrace the dimension F ($\equiv$ force) and Bridgman[5] gives an instructive example where analysis is carried out in terms of the set $MLTV$, where V represents the dimension of volume. Huntley[34] and Focken[25], in a large number of examples, successfully break down the dimension length, L, into orthogonal component dimensions X, Y and Z. Such techniques often lead to an impressive gain in the sharpness of the analysis.

It is, however, by no means invariably the case that an increase in analytical power does result. With disconcerting frequency it happens that when we work, for example, with the set $MXYZT$ — rather than with MLT — we find that, although the number of columns in our matrix has increased from 3 to 5, the rank nevertheless remains unchanged.

In our attempt to maximise r, it is clear, then, that it will be a waste of time to append to the indicial matrix any columns corresponding to new reference dimensions when these are dependent upon the pre-existing ones, for singularities would inevitably arise and no increase in rank would be attained.

Our approach, will be to examine the physical situation relating to a problem and to attempt to obtain clues that will suggest to us under what conditions the number of reference dimensions might be extended while still retaining their property of being independent or essential. We seek, in fact, to determine a criterion which will give us confidence that we are not extending the number of members in our chosen set of dimensions beyond some limitation implicit in the problem considered.

Pankhurst[52] gives an indication of what may be involved when he writes:

> The dimensions of force can be left unrelated to those of mass, length and time . . . and indeed should be in a problem of pure statics to which neither the law of acceleration nor the law of gravitational attraction applies. Similarly the dimensions of heat need not be identified with those of work, and indeed should not be, in a problem which does not involve the interconversion of thermal and mechanical energy.

Later in the same paper Pankhurst points out that it is legitimate to take X and Y as distinct dimensions of length when, as he says, 'what is happening in the vertical direction is independent of what is happening in the horizontal direction.'

These ideas point the way towards a systematic approach to the problem of maximising the number of members of the dimensional set, and we shall attempt to state with some degree of precision the requirements that have to be met.

5.2 Physical independence and physical completeness

If we are to increase the rank of an indicial matrix by appending a column which results from the introduction of a new reference quantity, then that column will have to be linearly independent of the pre-existing ones. We may, in consequence, expect that this independence will be reflected in the physical situation and that the new set of reference quantities will be independent in a purely physical sense which will in some way be associated with the mathematical independence. We use the term 'physical independence' to describe this and endeavour to make the concept clear.

Let us select a set of reference quantities which we propose to regard as complete, and let their dimensions be denoted by $ABCD$ .... The dimensional quantity A will then be 'physically independent' of the dimensional quantities

BCD . . . in some situation *S*, provided that this situation is such that:

1. *A* is not expressible in terms of *BCD* . . . *,

or, in a largely equivalent formulation,

2. a derived quantity, the dimensional representation of which includes *A*, is not expressible solely in terms of *BCD* . . . or in terms of derived quantities based solely upon *BCD* . . . .

Thus in the *MLT* system, and in most conventional situations, *M* is not expressible in terms of *L* and *T*, nor is any derived quantity whose representation includes *M* expressible in terms solely of *L* and *T* or of derived quantities based solely on *L* and *T*. We say, then, that *M* is physically independent of *L* and *T*. But in the system *VLT*, where *V* is the dimension of velocity, we have $V \equiv LT^{-1}$ and *V* is not, therefore, physically independent of *L* and *T*.

In intuitive terms, the physical independence of *A* implies that *A* is not related to, is not influenced by, does not transform into and does not react with *BCD* . . . or any quantities which may be dimensionally represented solely in terms of *BCD* . . . .

We say advisedly '*is* not expressible' and '*is* not related' rather than 'cannot'; and we also phrase our definition in such a way that physical independence is not an absolute characteristic but relevant to the given background to the situation *S*. We have, for instance, seen (4.3) that in the particular situation involving the equation *f* = *ma*, *L* cannot be considered as physically independent of *T*, although in most other situations it certainly is. Note also that although *A* may not be related to *BCD* . . . , it may and does enter into products with them, as *M* may and does enter into products with *L* and *T*.

It may be objected that there seems little difference between 'physical independence' and the 'formal' independence discussed in **1.7**. Admittedly the distinction is partly one of emphasis; but it goes further. In considering physical independence we seek to place ourselves in a position in which we can recognise independence directly from the physics of the situation. We are then able to proceed with the extension of the dimensional set with some assurance that no irritating singularity will arise in the indicial matrix. Were we to confine our attention to 'formal' independence only, we would render ourselves liable to be caught unawares by the occurrence of singularities and with no very clear idea of why they arise in particular cases.

Our approach will best be clarified by examples:

1. In the classical projectile situation of elementary mechanics, we denote the

---

* Where no confusion arises, we shall use the symbols *A* etc. to denote either the quantity *A* or its dimension.

dimension of the vertical direction by Y and that of the horizontal direction by X. Now since vertical acceleration ($g \equiv YT^{-2}$) is not related to or effected by the horizontal velocity ($u_x \equiv XT^{-1}$), and since no other variable containing Y in its dimensional representation is related to any other variable containing X in its representation, we may say that X is physically independent of Y.

But in a situation involving the tension in a string restraining a whirling bob, it is clear that acceleration in a radial direction is related to and dependent on the velocity of the bob in the tangential direction. In this second situation, then, we cannot treat R, the radial length dimension, as physically independent of Ψ, the tangential length dimension.

2. As stated by Pankhurst[52], force ($\equiv F$) will be physically independent of M, L and T in situations involving statics and unaccelerated motion. This follows directly from the fact that no resort is made to the equation $f = ma \equiv MLT^{-2}$. Where, however, explicit or implicit use is made of this equation, then clearly f will not be physically independent of MLT and no advantage will accrue from the consideration of F as a reference dimension in addition to MLT.

Note that Huntley[34] achieves an extension of the set MLT by distinguishing between the dimensions of M_q or 'mass regarded as quantity of matter' and M_i or 'mass in its inertial aspect'. This is legitimate and can be helpful in a number of problems, but reflection shows that the distinction is wholly equivalent to the introduction of F as a dimension physically independent of MLT and leads to no further increase in the sharpness of the analysis.

It may also happen, since force is a directional quantity, that force in the x direction, say, may be physically independent of force in the y direction. Using an obvious notation, the extension of the reference set to include F_x and F_y would then clearly be justifiable. In practice, however, no particular use of this approach will be made, since $F_x \equiv MXT^{-2}$ and $F_y \equiv MYT^{-2}$. We may, then, as readily work with a set involving extended length dimensions as with one involving extended force dimensions.

3. In a situation where q, the quantity of 'heat', is transformed into mechanical energy, H, the dimension of heat, will be physically dependent upon MLT. It follows that it will not be profitable to make use of the extended system $MLTH$. If, however, we are considering, say, the conduction of heat in a situation in which no transformation into mechanical energy takes place, then H may be regarded as physically independent of MLT and we may resort advantageously to the extended system.

4. In precisely the same way, Q, the dimension of electric charge, will be physically dependent upon MLT in problems where conversion takes place between electric and mechanical energy. No profitable use may then be made of the

extended $MLTQ$ system. But in problems where no such conversion takes place physical independence holds. Q, as it were, maintains its identity throughout the process considered, and the extended system may be used with advantage.

It will now be evident why physical independence, or the lack of it, is liable to be reflected in the relevant matrix. If, for example, the dimension A is to be physically independent of BCD . . . , in the sense which we have been discussing, then clearly there is no particular reason why the entries in the A column should be expressible as a linear combination of the entries in the BCD . . . columns. It follows that the sought-for increase in the rank of the indicial matrix will generally take place. While the term 'physical independence' refers primarily to dimensions, it will occasionally be useful to apply it to derived quantities, and no particular difficulties are found to arise as a result of this usage.

As a guide to the problem of deciding whether two dimensions in a set are or are not physically independent, it may be instructive to change the units in which we measure the quantities associated with those dimensions; provided physical independence holds, this may be done without effecting the validity of any result obtained.

An example illustrates this important point. A body moving in circular orbit has an acceleration directed towards the centre and physically dependent upon the velocity in the tangential direction. We have, in fact $a = v^2/r$. If now we treat the radial and tangential components of length as reference dimensions, and if we measure the former in feet and the latter in metres, we find that the magnitude of the acceleration, as calculated from the equation, becomes rather difficult to interpret. In fact, if dimensional homogeneity is to be maintained, the units of acceleration will be $m^2 \, ft^{-1} \, s^{-2}$, and it is unacceptable that the tangential length unit of the metre should find any place in the unit of radial acceleration.

Contrast this with the case of the projectile problem where physical independence exists between the two orthogonal length dimensions. Here we may measure vertical heights in feet and horizontal lengths in metres, should we so wish, with no confusion arising and with no quantity appearing which is measurable in units defined partly in terms of feet and partly in terms of metres.

This possibility of making a differential change in the units of physically independent quantities is associated with the justification for introducing differential changes of scale and other types of distortion in models. It follows that the concept of physical independence has an important bearing on model theory; this will be considered in Chapter 10.

In the foregoing we have intentionally placed an emphasis upon the role of physical independence. We recall now from 1.7 that the criteria for a complete set of dimensions are two, involving not only independence but also the possibility of deriving from the members of the set all other quantities which enter into the situation. As an analogue of this we will also consider 'physical completeness', and with this idea in mind we now say:

A set of reference quantities is 'physically complete' when the background situation is such that each member of the set is physically independent of the others and when each further quantity relating to the situation, and not included in the set, is physically dependent upon members of the set.

It will be clear that optimum extension of the reference set occurs when this set is physically complete. This follows from the two observations:

1. No further extension of the reference set is required, for it is entailed by the definition of physical completeness that any additional reference quantities will be physically dependent upon pre-existing members of the set and will, therefore, serve no useful purpose.

2. No reduction in the set is possible without it losing the property of physical completeness, for there would then exist a quantity outside the set which was independent of its members and could not, therefore, be derived from them.

Once again, physical completeness is not an absolute characteristic but varies with the situation considered. Thus, in a situation where XYZ were physically independent of one another, MLT could not be regarded as a complete set in that a further extension would be possible.

There has been a persistent tendency for writers on dimensional analysis to resist any extension of the number of members of the set of reference dimensions. There was, for example, in 1915 a famous and much quoted discussion between Lord Rayleigh[58] and Dr Riabouchinsky[59]. The latter pointed out that an equation for heat flow derived by Rayleigh on the assumption that M, L, T and Θ (temperature) were fundamental and basically independent quantities would have been different and less useful if temperature had been expressed in terms of MLT. Lord Rayleigh remarked that:

It would indeed be a paradox if further knowledge of the nature of heat afforded us by the molecular theory puts us in a worse position than before in dealing with a particular problem.

It should, however, by this time be evident that although an advance in knowledge may enable us to express one quantity, hitherto regarded as independent, in terms of other quantities, it will nevertheless be an advantage from the point of view of dimensional analysis to continue to regard that quantity as a reference dimension – unless the situation under consideration actually involves its physical dependence upon the others. However desirable it may be from the standpoint of conventional physics to relate quantities to their most fundamental components, this is not necessarily the task of dimensional analysis, where we have always to bear in mind that the precision of a solution increases with the rank of the indicial matrix. It will, then, be our aim to carry out the analysis of a problem in terms of a set of reference dimensions that satisfies the criteria of 'physical completeness' in the context of the relevant background situation.

The successful use of the dimensional approach involves, not the blind application of a rule of thumb, but a sensitive physical intuition. It will already have been accepted that this intuition is required in the selection of the set of variables that are considered relevant to a problem. It now becomes apparent that this intuition is equally pertinent to the selection of the set of reference dimensions with which we are to operate and to the decision of how best this set may be extended to ensure physical completion.

5.3   Extensions to the length dimension

The replacement of the single length dimension, L, by three orthogonal components is one of the more valuable techniques to be discussed. The extension will frequently be in terms of XYZ, representing the dimensions of the component length quantities along the three cartesian axes. Depending upon the nature of the situation under analysis, however, the extension may also be made in terms of $R\Psi Z$, representing dimensions corresponding to length components in a radial, a tangential and in an axial direction. Again, the extension may be in terms of intrinsic dimensions specifically designed to correspond to some natural framework into which the phenomenon being treated happens to fit. We give examples of each of these approaches.

Meanwhile we would warn that an uncritical and over-enthusiastic use of extensions to the length dimension, in cases where physical independence does not in fact occur, may frequently lead to inconsistency (5.5). Consequent upon this difficulty, a number of attempts have been made to resolve it by the introduction of vectorial methods. Gessler[29], however, refers to certain doubtful consequences that are liable to arise from this technique. It would, for instance, seem to be implied that, if the length dimension be regarded as a vector, then an area lying in the xy plane should presumably be expressed as a vector product $X \times Y$ and will, therefore, be a vector lying in the z direction, which hardly seems appropriate. He also feels that it is straining matters if certain essentially scalar quantities, such as viscosity, the dimensions of which include length, are to appear with a representation that involves a vector.

Gessler is, nevertheless, reluctant wholly to abandon the vectorial approach and proposes the adoption of a 'mixed' $MLTI_xI_yI_z$ system, where L is the dimension of scalar length and $I_xI_yI_z$ are the 'dimensions' of the unit vectors in the x, y and z directions. His development appears free of inconsistency and has the advantage of facilitating the avoidance of error. In particular it systematises, but is by no means essential to, the consistent representation of certain quantities the dimensions of which may not be too obvious when expressed wholly in scalar terms. Nevertheless, the use of the 'vectorial' dimension leads to no increase in the information provided by the analysis and, in any case, it represents a concept which would have to be rather carefully defined. It certainly would not, for example, be accommodated by the accepted definition of a dimension due to Maxwell (1.5). Our feeling is, then, that Gessler's admittedly ingenious system is somewhat inelegant and has no advantages to offer over the more natural methods in this book.

In our own thinking, we emphasise that XYZ are not vectors. They are essentially the dimensions of the scalar length components lying in the x, y and z directions. Moreover we are not impressed by such inconsistencies as are occasionally reported in the literature, for these invariably result from the use of the extended length system in conditions where physical independence does not exist (5.5). Provided the use of the $MXYZT$ system be restricted to the conditions for which it is formulated and where the length components are not physically dependent upon one another, no difficulties are experienced.

Let us, then, proceed with the consideration of a number of examples. In general we find that the greater the degree of asymmetry in the background situation, the greater will be the value of working with an extended system. If, indeed, there is complete asymmetry in the three orthogonal directions, say x, y and z, then each of the corresponding columns in the indicial matrix will contain different entries and these will frequently be linearly independent of one another. The rank will, therefore, be greater by 2 than is the rank of the matrix based on the conventional MLT system and the number of DPs in a compelte set will be fewer by 2.

Where, as frequently happens, two of the directions in a problem are symmetrical (as in axisymmetric situations) the rank of the matrix will be increased by one only. This is also liable to be the position where the phenomenon considered takes place wholly in a plane and may, therefore, be adequately described in terms of, say, X and Y only.

Finally, where there is complete symmetry, the rows corresponding to X, Y and Z will be identical; no physical independence exists and no advantage over the MLT system will be obtained by working in $MXYZT$.

The position may be illustrated by two quite trivial examples. The case of complete symmetry is applicable to the equation $m = \rho v$, relating mass to density and volume. Considered in terms of the MLT system, the indicial matrix is:

	M	L	T
m	1	0	0
ρ	1	−3	0
v	0	3	0

This, being of rank 2 and related to 3 variables, yields $(3 - 2) = 1$ DP, namely $(\rho v/m)$. The corresponding matrix in the $MXYZT$ system would be

	M	X	Y	Z	T
m	1	0	0	0	0
ρ	1	-1	-1	-1	0
v	0	1	1	1	0

Due to the fully symmetrical situation, there is no physical independence between X, Y and Z; the associated columns in the matrix are identical and the rank remains equal to 2. Nothing, therefore, is gained by our decision to work in the extended system. (Note also that, in this particular case, nothing is lost; our one DP is still thrown up and no inconsistency arises.)

In contrast with this example, consider the fully asymmetric position where we determine the weight W of a rectangular block of density ρ and with sides of length a, b and c. In the MLT system we should be faced with a matrix of rank 3 and, consequently, $(6 - 3) = 3$ DPs, namely

$$(W/\rho g a^3), (a/b) \text{ and } (a/c)$$

In the $MXYZT$ system, however, we have, taking c as lying in the vertical or z direction, the matrix

	M	X	Y	Z	T
W	1	0	0	1	-2
ρ	1	-1	-1	-1	0
a	0	1	0	0	0
b	0	0	1	0	0
c	0	0	0	1	0
g	0	0	0	1	-2

Note here that the gravitational acceleration g is effective in the z direction and consequently has the dimensions ZT^{-2}. The matrix is of rank 5 and accordingly there is $(6 - 5) = 1$ DP only, that is $(W/\rho abcg)$, instead of the 3 DPs obtained when we failed to take advantage of the asymmetry of the situation and confined our attention to the unextended system MLT.

Observe how, in this example, the contribution to the overall weight made, say, by the side of length a, lying in the x direction, is in no way related to or dependent upon the contribution made by the side of length b lying in the y direction. This physical independence is reflected in the algebraic independence of the columns of the matrix. We also confirm that different units may be used for the three independent length dimensions without obtaining any inconsistency in the result. Our one DP yields the equation $W = k.\rho abcg$ and, inserting in this the units of, say, inches and feet for the two horizontal length directions and metres for the vertical ones, we obtain as our units of weight:

$$\frac{kg}{in\ ft\ m} \times in\ ft\ m \times \frac{m}{s^2} = kg\ m/s^2$$

5.3.1 Application of a couple to a prism

We now set out a number of less trivial examples. The first of these considers another fully asymmetric situation and we investigate the angle through which a right rectangular prism is twisted as a result of the application of a couple. We take this angle α as the dependent variable which is effected by the length of the prism l, its thickness a, its width b, the applied couple C and the modulus of rigidity η. The indicial matrix is then

	M	X	Y	Z	T
α	0	0	0	0	0
l	0	0	0	1	0
a	0	1	0	0	0
b	0	0	1	0	0
C	1	1	1	0	-2
η	1	-1	-1	1	-2

As required by the conditions for physical independence, no quantity whose dimensional representation includes X is expressible in terms of quantities based on Y and Z. Similarly for Y and for Z. The dimensional structures themselves appear straightforward, except possibly in the cases of C and η. Referring to figure 6, we see that the couple C, defined as the magnitude of a pair of parallel and opposite forces multiplied by the perpendicular distance between their lines

Figure 6

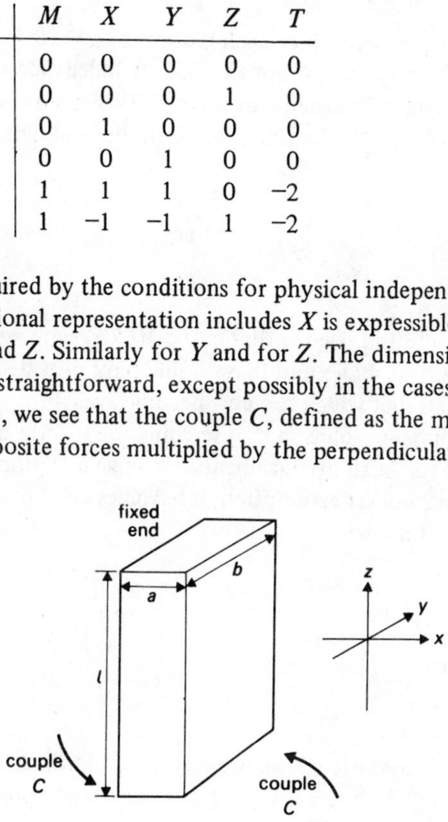

of action, must always have components whose dimensions are

$$(MX/T^2)Y \quad \text{and} \quad (MY/T^2)X$$

Since these two forms are equivalent, we have no difficulty in arriving at the representation $C \equiv MXYT^{-2}$ shown in the matrix. Secondly, the modulus of rigidity is defined as

$$\eta = \frac{\text{tangential force}}{\text{area}} \div \text{strain}$$

Now the component dimensions of the tangential force will be MXT^{-2} and MYT^{-2}. The area will in each case have the dimension XY, while the strain in the two cases will be respectively represented by XZ^{-1} and YZ^{-1}. We have, then, as our two components,

$$\eta \equiv \frac{MX}{T^2 XY} \div \frac{X}{Z} \quad \text{and} \quad \eta \equiv \frac{MY}{T^2 XY} \div \frac{Y}{Z}$$

These two forms are once again equivalent, since each reduces to $\eta \equiv MZX^{-1}Y^{-1}T^{-2}$.

With regard to the matrix itself, note that although XYZ are independent, as is to be expected, it happens that M is linearly dependent upon T. This results in a singularity and the rank is 4, giving $(6-4) = 2$ DPs as contained in the equation

$$f\left(\frac{Cl}{\eta a^2 b^2}, \alpha\right) = 0$$

or $\quad C = \dfrac{\eta a^2 b^2}{l} \phi(\alpha)$

Although this is not a complete solution, we see that in the MLT system we should have been left with as many as 4 DPs and these would have provided considerably less information than that which has now become available.

We may, however, obtain a complete solution if, discarding the strictly dimensional approach, we assume that for small displacements the angle α is proportional to the applied couple C. With this modest assumption, it becomes clear that the final solution to our problem can only be

$$\alpha = k \cdot \frac{Cl}{\eta a^2 b^2}$$

where k is a numerical coefficient.

5.3.2 Range of a projectile

An instructive example considers the range R of a projectile fired horizontally with velocity v from a point situated at a height h above a plane. As already pointed out, motion in the horizontal x direction is independent of motion in the vertical y direction and we may write

	X	Y	T
R	1	0	0
u	1	0	−1
g	0	1	−2
h	0	1	0

This yields at once the required solution:

$$R = ku(h/g)^{1/2}$$

Note how any change in the units of vertical length that appears in the structure of g and h will leave the value of R unaffected.

Consider next the position where the projectile is fired from ground level at an angle of α from the horizontal. In this case we have two (independent) velocity components u_x and u_y, where $\tan \alpha = u_y/u_x$. Analysis immediately gives the solution

$$R = k(u_x u_y/g)$$

If, however, we attempt to generalise the case by considering a projectile fired at an angle of α from a height h above the plane, we are left with $(5 - 3) = 2$ DPs, namely

$$(u_x u_y/Rg) \quad \text{and} \quad (u_y{}^2/gh)$$

which make possible only the incomplete solution

$$R = (u_x u_y/g)\, \phi\, (u_y{}^2/gh)$$

This lack of success results directly from the fact that conventional analysis shows that the correct result is of a nature which cannot be put in the form of a PP.

5.3.3   Energy of a vibrating wire

We now re-examine the example of **4.3**, dealing with the energy of a vibrating wire, and we show how the partial solution previously obtained may be rendered complete by working in the extended $MXYT$ system.

Define the x direction as lying along the axis of the wire and the y direction as lying perpendicular to x and within the plane in which the vibrations take place. It follows that no force or motion occurs in the z direction, and, in consequence, no quantity arises which requires to be defined in terms of the dimension Z. The dimension of the wire length will be simply X; the linear density of the wire will be MX^{-1}; the amplitude of vibration, lying in the y direction, will have the dimension Y; the energy of vibration, being due to motion in the y direction only, will be represented by $E \equiv MY^2 T^{-2}$; and finally the tension, being effective along the line of the wire, must be represented by $f \equiv MXT^{-2}$. The matrix of **4.3** may, then, be rewritten

	M	X	Y	T
E	1	0	2	-2
l	0	1	0	0
ρ	1	-1	0	0
A	0	0	1	0
f	1	1	0	-2

The rank is now 4, whereas the rank of the matrix in **4.3** was only 3. The number of DPs, therefore, reduces to $(5 - 4) = 1$ and a complete solution becomes possible. Writing out the indicial equations and solving gives $\pi_1 = (El/A^2f)$, leading to

$$E = k \cdot A^2 f/l$$

It follows from this that the function ϕ, left undetermined in **4.3**, is now seen to be simply $\phi = (l/A)^{-1}$.

Here, as in most cases, we may eliminate the tedium of solving the indicial equations by writing our one DP directly from an inspection of the matrix. This is the method indicated in **4.2** and, in view of its common application, we set out the detailed working as applied to the present example. In the following schema, the dimensions shown in the right-hand column represents the dimensions of π_1 as progressively revealed during the course of the step-wise procedure indicated:

Suppose first that E appears in the numerator of π_1:

$$E \equiv \frac{MY^2}{T^2}$$

If the dimension of T in the product is to be zero, then f must appear in the denominator:

$$\frac{E}{f} \equiv \frac{MY^2 T^2}{T^2 MX} \equiv \frac{Y^2}{X}$$

If, now, the dimension of Y is to be zero, A^2 must appear in the denominator:

$$\frac{E}{A^2 f} \equiv \frac{MY^2 T^2}{T^2 MXY^2} \equiv X^{-1}$$

Finally, if the dimension of X is to be zero, then l will appear in the numerator and we have:

$$\frac{El}{A^2 f} \equiv \frac{MY^2 T^2 X}{T^2 MXY^2} \equiv [1]$$

and $(El/A^2 f)$ is the required DP.

5.3.4 Flow over a broad-crested weir

We introduce an example in which the co-ordinate system is designed to suit the particular situation considered and we take up the problem of flow over a broad-crested weir (figure 7). Were we to work in the XYZ system, we should invite

Figure 7

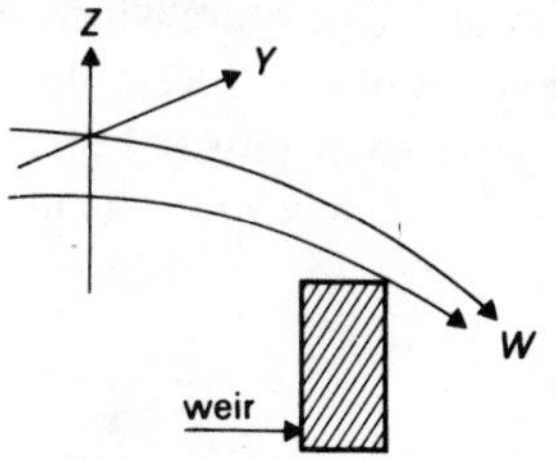

difficulty in view of the fact that the flow is partly in the horizontal and partly in the vertical direction. We consider, therefore, the possibility of working with three length dimensions, WYZ, with W being directed along the stream lines, Y along the line of the weir and Z representing the vertical dimension. W and Z, however, will not be orthogonal and cannot be regarded as physically independent of one another, so we abandon any attempt to distinguish them and subsume them under the one dimension L. Y, then remains physically independent of L and we may legitimately treat it as an independent reference dimension.

Taking the dependent variable $\dot{m}$ as the mass flowing per unit time and the independent variables as the density of water ρ, the head above the weir h, the breadth of weir b and the acceleration due to gravity g, we have the matrix

	M	L	Y	T
$\dot{m}$	1	0	0	−1
ρ	1	−2	−1	0
h	0	1	0	0
b	0	0	1	0
g	0	1	0	−2

This yields one DP and the solution is

$$\dot{m} = k \cdot p\, g^{1/2} h^{3/2}\, b$$

5.3.5 Expansion of a bimetallic strip

An example, at first sight deceptive, considers a bimetallic strip. On increasing the temperature by $\Delta\, \theta^{0}$, this deforms into a circular arc and we require to find the angle α which this arc subtends.

Let the difference in the coefficients of expansion of the two metals be β; let the thickness of the strip be h and its length be l. Since the dimensions M and T are not required for the representation of any of the significant quantities, we work simply with the system $LR\Theta$, L being the dimension of length along the axis of the strip, R being the dimension of length orthogonal to L and passing through the centre of the circle and Θ being the dimension of temperature which, at this stage, we may see is a quantity physically independent of L and R, in that the situation considered involves no transformation between quantities expressed in terms of L and R and those expressed in terms of Θ. We now write the indicial matrix

	L	R	Θ
α	1	−1	0
l	1	0	0
h	0	1	0
β	0	0	−1
$\Delta\theta$	0	0	1

Here the dimensions of β are in accordance with the definition of this quantity in terms of the ratio 'strain' divided by unit temperature increase. The matrix throws up $(5 - 3) = 2$ DPs which yield the equation

$$\alpha = (l/h)\,\phi(\beta\Delta\theta)$$

Although this is as far as we can go, restricting ourselves to dimensional methods, we may reasonably assume that $\alpha \propto \Delta\theta$, and making use of this additional relationship we obtain the complete solution:

$$\alpha = k \,.\, (l/h)\,\beta\Delta\theta$$

which may be confirmed by conventional analysis, when we find that the undetermined coefficient $k = 1$.

Commenting upon this problem we make a possibly subtle point. As is apparent from the matrix, L and R are independent of one another, and it may be advisable to comment upon the physical interpretation of this. Despite the algebraic independence of these two dimensions, we may yet have a feeling that events happening in the R direction (i.e. changes in the radius of curvature) seem in some sense to be dependent upon events in the L direction (i.e. differential expansion along the strip axis). Upon closer examination of the situation, however, it becomes clear that changes in the radius of curvature are physically dependent upon h only. In fact, if we work with the radius of curvature ρ as our dependent variable, rather than with the angle subtended, we find $\rho = h \,.\, \beta\Delta\theta$. Here the length l drops out of the equation and there is no dependence of ρ ($\equiv R$) upon any quantity with a dimensional representation that includes L. That we choose to regard the dependent variable as α rather than ρ will in no way alter our finding that R and L are, in fact, physically independent dimensions.

5.4 The principle of symmetry

A problem frequently arises where more than one representation of a derived quantity becomes possible in the $MXYZT$ system. How, for example, are we to express the dimensions of the radius of a circle lying in the xy plane?

There are two approaches here. We first observe that, since there is no preferred direction, we may write $r \equiv X, r \equiv Y$ or $r \equiv X^a Y^b$ for any a, b such that $a + b = 1$. Now, provided that the situation is symmetrical with respect to the x and y directions, there is no reason to give more weight to one direction than to the other and, to maintain this symmetry, we may write the dimensions of r as $X^{1/2} Y^{1/2}$.

A slightly different approach would be to consider the area of the circle, rather than its radius, as the significant variable. This leads to $A \equiv XY$. We may then, if we wish r to appear explicitly in the solution, consider it as being proportional to the square root of A, thus returning to the dimensional equivalence $r \equiv X^{1/2} Y^{1/2}$.

With these considerations in mind, we enunciate the following 'principle of symmetry':

> Where the dimensional representation of a quantity varies with its orientation and provided there is no preferred orientation entailed by the background situation, then we may work with a symmetric representation of that quantity, giving an equal weight to each of the representations corresponding to the orthogonal orientations concerned.

In order more rigorously to justify this principle, we argue as follows. In the MLT system, let the length dimensions of a quantity A be L^n. Then in the $MXYZT$ system, the component dimensions of A may be written according to the directions in which it is considered as either:

$$A_1 \equiv X^a Y^b Z^c \equiv L^n$$
$$A_2 \equiv X^d Y^e Z^f \equiv L^n$$
$$A_3 \equiv X^g Y^h Z^i \equiv L^n$$

where $a + b + c = d + e + f = g + h + i = n$. Now, unless there is any preferred orientation, we are justified in arguing that if A is a significant quantity so also will be the product $A_1 . A_2 . A_3 = A^3$. Furthermore, there is no objection if, for convenience, we prefer to work not with A^3 but with

$$(A^3)^{1/3} = A = X^{(a+d+g)/3}\, Y^{(b+e+h)/3}\, Z^{(c+f+i)/3} = X^{n/3} Y^{n/3} Z^{n/3}$$

which confirms the principle of symmetry.

This principle is a useful working tool. It is, however, generally possible to avoid an appeal to it by so changing the co-ordinate system that it fits the background more neatly and naturally than does the cartesian. Thus our initial difficulty in finding the dimensional representation of a circle lying in the xy plane vanishes as soon as we use polar co-ordinates. The length dimensions will then be denoted by R and Ψ, and the dimensions of r will be simply R. Alternatively, if we are considering an axisymmetric situation, it may be simplest to regard the axial length dimension as represented by X, while any lengths lying in the yz plane, irrespective of their orientation, may be represented by L.

Where there is no preferred direction and where resort is made to the principle, it generally happens that, say, the X and Y columns of the indicial matrix will be identical. There will, therefore, be dependence between them and no advantage is obtained from the extended system. It will, then, be appreciated that the principle of symmetry is not an essential tool of dimensional analysis; its use is confined to the convenience which it affords in carrying out routine work.

If the magnitude of a variable is a function of its orientation, the principle of symmetry is not applicable and it should not be necessary to warn against a possible temptation to 'weight' the dimensions of a quantity in such a way that the orientation may appear to be reflected in the representation adopted. Thus in the Pythagorean equation $l^2 = x^2 + y^2$ we cannot put $l \equiv X^a Y^b$ with $a + b = 2$ and weighted to indicate the gradient of l. Indeed, this equation represents a typical case in which lack of symmetry coupled with physical dependence between length dimensions results in their being a lack of dimensional homogeneity in the extended $MXYT$ system — although, of course, there is full homogeneity in MLT.

5.4.1   Rise in a capillary tube

An example illustrative of the principle of symmetry considers the height h to which a liquid of density ρ and surface tension τ rises in a capillary tube of radius r. Taking z as the vertical direction and x and y as two perpendicular horizontal directions, we prepare the matrix

	M	X	Y	Z	T
h	0	0	0	1	0
ρ	1	-1	-1	-1	0
r	0	$\frac{1}{2}$	$\frac{1}{2}$	0	0
g	0	0	0	1	-2
τ_z	1	$-\frac{1}{2}$	$-\frac{1}{2}$	1	-2

This is straightforward enough. With regard to the dimensions of surface tension, we recall that this quantity is defined as 'force per unit length exerted in a plane tangential to the surface and in a direction normal to the unit length considered'. It follows that τ_z, the vertical component of surface tension, may be represented either by $MX^{-1}ZT^{-2}$ or by $MY^{-1}ZT^{-2}$. Applying the principle of symmetry, we allot equal weights to both x and y directions to obtain the equivalence $\tau_z \equiv MX^{-1/2}Y^{-1/2}ZT^{-2}$. (Should we prefer to define surface tension in terms of 'free surface energy', then the dimensions will be unchanged since force/length is dimensionally equivalent to energy/area.)

The representation $r \equiv X^{1/2}Y^{1/2}$ needs no comment.

We have, then $n = 5$ variables expressed in terms of 5 reference dimensions. The entries in the X and Y columns are, however, identical, as is to be expected since the fact that we are able to make use of the principle of symmetry in this problem indicates that no physical independence occurs between X and Y. It follows that the matrix is singular and of rank $r = 4$. Determining the one DP by inspection we find $\pi_1 = (h\rho r g/\tau_z)$. As a refinement we now introduce θ as the angle of contact between liquid and tube wall, giving $\tau_z = \tau \cos \theta$, and the final result becomes

$$h = k(\tau \cos \theta / \rho r g)$$

As already pointed out, we could have avoided an appeal to the principle of symmetry simply by working with the dimension L representative of any length in the xy plane, thus reducing the number of columns in the matrix to 4.

Alternatively, we may use a more natural framework and rework the problem in cylindrical co-ordinates, writing the components of the length dimension in the radial and tangential directions as R and Ψ. The component of the axial (vertical) direction remains Z and the indicial matrix becomes

	M	R	Ψ	Z	T
h	0	0	0	1	0
ρ	1	-1	-1	-1	0
r	0	1	0	0	0
g	0	0	0	1	-2
τ_z	1	0	-1	1	-2

This leads directly to the result previously obtained.

There is, however, a point of interest. Since the solution is known to involve one DP only, it follows that we cannot have increased the rank of the matrix above the previous value of 4, even though the columns corresponding to the three length components are now seen to be independent. This implies that some other relationship of independence must have arisen and, indeed, inspection shows that whereas previously X was dependent upon Y, we now have M dependent upon Ψ.

Our example confirms that the principle of symmetry may be used as a matter of convenience only and not of necessity. No gain in information derives from its use, since the assumption of symmetry between two or more length dimensions entails that there is a linear dependence between those dimensions and the necessary condition for increasing the rank of the indicial matrix is, therefore, absent.

5.5   Sources of difficulty where physical dependence persists

At this point we consider in more detail the question of what happens when an 'illegitimate' addition is made to the set of dimensions – when, that is to say, we attempt to work with an extended set under conditions in which physical independence does not occur. In such cases the solution obtained need not, in general, be erroneous. Where, for example, the additional reference dimension is a form of energy, the attempt to work with it, when it is in fact dependent upon other members of the set, will normally result merely in the introduction into the matrix of a singularity and, although no information is gained, nothing has been lost.

Where the additional reference dimensions are components of the length dimension, the position is liable to be more subtle. The use of the principle of symmetry is liable to result only in the introduction of a harmless singularity, as in **5.4.1**. Alternatively, we may be faced with a complete breakdown in homogeneity as was the case with the Pythagorean equation, mentioned in **5.4**.

As a simple illustration of this difficulty, consider the acceleration of a particle moving in a circle, this being given by $a = r\omega^2$. Now write $v = r\omega$ and transform the equation into the form $a = v^2/r$. If we examine this latter result in terms of the orthogonal length dimensions R and Ψ we shall be faced with an apparent collapse of the principle of dimensional homogeneity in that

$$a \equiv \frac{R}{T^2} \not\equiv \frac{\Psi^2}{RT^2} \equiv \frac{v^2}{r}$$

This lack of equivalence results directly from the substitution of v for $r\omega$. In terms of the dimensions $R\Psi$, these two quantities, although numerically equal, are not dimensionally homogeneous since $v \equiv \Psi/T$ while $r\omega \equiv R/T$. The position is, in fact, similar to that which we faced in 2.3.1 where we substituted kp for ρ. Just as we previously had to consider k as a dimensional constant, so now we have to introduce a dimensional 'transformation coefficient' of magnitude equal to unity and of dimensions $k \equiv \Psi/R$. The substitution $v = k \cdot r\omega$ may then be validly made and we run into no danger of breaking the rule which involves equating like quantities with like.

It follows that a definitive criterion for the 'legitimate' extension of a set of reference dimensions may be based upon an examination of the fundamental equation descriptive of the situation under consideration. In general this may not be a particularly helpful approach, since, once the equation be known, dimensional analysis may have little or no significant contribution to make in finding a solution. It is, nevertheless, often of interest to obtain an understanding of why we can, or cannot, successfully work with an extended dimensional set in particular cases.

Again, it may be that, although the basic equations are known, we seek information concerning some particular solution. This happens in 7.2 where we examine the Navier–Stokes equations and we ascertain precisely under what conditions these equations are homogeneous in an extended system and under what conditions, therefore, that extended system may profitably be used.

Further to clarify the position, we shall in this section examine two examples of equations which are homogeneous in MLT but apparently heterogeneous in $MXYZT$ and we shall have the satisfaction of seeing just why the extended system breaks down.

5.5.1   Pressure beneath the earth's surface

Consider the lateral pressure p developed at depth z beneath the earth's surface as a result of the weight of the overlying rock, which we take as having a density ρ and Poisson's ratio ν. This lateral pressure may be shown by conventional methods to be given by

$$p = \frac{\nu}{1 - \nu} \cdot \rho g z \tag{1}$$

With z as the vertical direction and x and y as two orthogonal horizontal directions, an application of the principle of symmetry would give the left-hand side of equation 1 as $p \equiv MZ^{-1}T^{-2}$ while on the right we would have $\rho g z \equiv MX^{-1}Y^{-1}ZT^{-2}$. In consequence, the dimensions of the factor $\nu/(1 - \nu)$ would appear to be given by XYZ^{-2}. It will be objected that $\nu/(1 - \nu)$ cannot be dimensional, for this factor can have meaning only if ν is a pure number and, in the $MXYZT$ system, ν being the ratio of orthogonal strains, cannot be a numeric.

Once again, the difficulty arises from the algebraic summation of quantities which are dimensionally equivalent in the MLT system, but which are not so when L is resolved into components. In more detail, equation 1 derives from an equation in the theory of elasticity:

$$\epsilon_x = \frac{\sigma_x - \nu(\sigma_y + \sigma_z)}{E} \tag{2a}$$

where ϵ_x is the strain in the x direction, σ_x etc. are the normal stresses in the x etc. directions, ν and E being Poisson's ratio and Young's modulus respectively. If we rewrite this in terms of component dimensions of length we have:

$$\epsilon_x = \frac{\sigma_x - \nu_y \sigma_y - \nu_z \sigma_z}{E} \tag{2b}$$

where $\nu_y \equiv X^2/Y^2$ and $\nu_z \equiv X^2/Z^2$.

These alternative dimensional forms of Poisson's ratio emphasise that two different quantities are being encountered and it is only their numerical equality in an isotropic solid that enabled us to put ν in place of ν_y and ν_z outside the bracket of equation 2a. We also note, by considering the dimensional representation of these quantities in $MXYZT$, that they represent ratios of stresses rather than of strains as we had at first supposed.

With this in mind we review the conventional development of equation 1. Let σ_x represent the lateral stress induced in the rock of the earth's crust as a result of the superincumbent weight. We then argue that since, in an isotropic medium, there is no reason why displacement should take place in one horizontal direction rather than in another, we may put ϵ_x in equation 2a as equal to zero. Still taking z as the vertical direction (figure 8), we also put the lateral stress σ_y equal to σ_x, giving

$$\sigma_x - \nu(\sigma_x + \sigma_z) = 0 \tag{3}$$

or $\sigma_x = \dfrac{\nu}{1 - \nu} \cdot \sigma_z$

which is equivalent to equation 1.

Figure 8

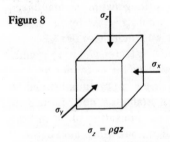

$\sigma_z = \rho g z$

The reason that this is dimensionally heterogeneous in the $MXYZT$ system is now evident, for we should, from the start, have based our argument upon equation 2b, the dimensionally correct equation in the extended system, rather than on equation 2a which is correct only in the MLT system. Equation 2b, it will be observed, yields the valid analogue to equation 3, which is

$$\sigma_x - \nu_y \sigma_y - \nu_z \sigma_z = 0$$

Summarising, it is clear that what is happening in the x and y directions is not independent of what is happening in the z direction. XYZ are not, therefore, physically independent and it is for this reason that our original naive attempt to demonstrate dimensional homogeneity in equation 1 broke down.

5.5.2   Deflection of a rectangular beam

As a second example of this type of difficulty, consider the maximum deflection of a rectangular beam supported at its ends. Conventional analysis gives

$$d = mgl^3/Ebd^3$$

where m is the total mass, l is the length, b and d are respectively the beam breadth and thickness, and E is the Young's modulus. A rapid check shows that this equation is dimensionally homogeneous in MLT. But in $MXYZT$, with z as the vertical direction and x the direction of the beam axis, we have

$$d \equiv Z \not\equiv Z(X^2/Z^2) \equiv mgl^3/Ebd^3$$

As an infallible method for determining what is wrong, we again review the conventional development of the equation, checking the homogeneity at each step. At one stage we have $1/\rho = m/EI$, where ρ is the radius of curvature and I the moment of inertia. This is homogeneous in $MXYZT$. The next step, however, is the substitution of d^2z/dx^2 for $1/\rho$, which is where the dimensional equivalence breaks down since:

$$\frac{d^2z}{dx^2} \equiv \frac{Z}{X^2} \not\equiv \frac{1}{Z} \equiv \frac{1}{\rho}$$

Once again, then, homogeneity has been lost by the substitution of numerically equal but dimensionally heterogeneous quantities and this situation has, moreover, arisen only as a result of a lack of physical independence between X and Z.

We may, in fact, say quite generally that where an equation is homogeneous in MLT but not in $MXYZT$, this inevitably implies that there is no physical independence between X, Y and Z and homogeneity may, then, be restored only by the introduction of dimensional constants.

Costa[11], although he makes no explicit reference to the idea of physical independence, is conscious of this situation when he writes:

The difficulties posed by the dimensional constants in dimensional analysis will increase with the consideration of directions in directional analysis because some equations that were considered as complete in dimensional analysis will cease to be so in directional analysis. In fact in directional analysis new dimensional constants as well as directional constants are to be considered.

(Here 'directional analysis' is Costa's term for 'dimensional analysis involving orthogonal components of the length dimension' and the word 'complete' refers to what we have described as dimensional homogeneity.)

5.6   Alternative approaches to Poiseuille's problem

It can happen that the introduction of additional members of the reference set may yield an unexpected bonus; and it will be profitable to consider various approaches to the analysis of Poiseuille's problem of laminar flow through a tube, a problem, incidentally, to which repeated reference will be made in succeeding pages.

As a preliminary we derive the dimensions of viscosity in an extended system and note that this representation may only be used where the validity of that system is justified. (See also 7.2.) Recall first that (dynamic) viscosity is defined in terms of the equation

$$\tau = \mu dv/dy$$

Here τ is the shear stress set up in the fluid which we take to be flowing in the x direction. The (orthogonal) direction of shear will be y, while the direction normal to the xy plane will be z. We then have

$$\tau \equiv \frac{MXT^{-2}}{XZ} \equiv MZ^{-1}T^{-2} \qquad v \equiv XT^{-1} \qquad y \equiv Y$$

and, upon substitution,

$$\mu \equiv MZ^{-1}T^{-2} \cdot \frac{Y}{XT^{-1}} \equiv MX^{-1}YZ^{-1}T^{-1}$$

We also note that it follows from this that kinematic viscosity, defined as $\nu = \mu/\rho$, has the dimensions

$$\nu \equiv \frac{MX^{-1}YZ^{-1}T^{-1}}{MX^{-1}Y^{-1}Z^{-1}}$$

$$\equiv Y^2 T^{-1}$$

Poiseuille's problem may now be presented in a variety of ways, but we initially consider the mean fluid velocity U as the dependent variable while the independent variables are the pressure gradient dp/dx, the density of the fluid ρ, the viscosity of the fluid μ, and the tube radius r. We start by carrying out the analysis in terms of MLT and obtain the matrix

	M	L	T
U	0	1	−1
dp/dx	1	−2	−2
ρ	1	−3	0
μ	1	−1	−1
r	0	1	0

With 5 variables and a rank of 3 we have in consequence 2 DPs and an incomplete solution. We also note that had we considered the mass or volume flow rate, $\dot{m}$ or Q, as the dependent variable, rather than U, the outcome would have been similar.

We decide, therefore, to take advantage of the physical independence that exists between the direction of flow and directions in the yz plane. Working in the $MXYZT$ system and using the principle of symmetry, we have the matrix

	M	X	Y	Z	T
U	0	1	0	0	−1
dp/dx	1	0	−1	−1	−2
ρ	1	−1	−1	−1	0
μ	1	−1	0	0	−1
r	0	0	$\frac{1}{2}$	$\frac{1}{2}$	0

As is to be expected, the columns corresponding to Y and Z are identical, but we have, nevertheless, increased the rank to 4 and we have, therefore, a complete solution:

$$U = k \cdot \frac{r^2}{\mu} \cdot \frac{dp}{dx} \tag{1}$$

Had we worked with $\dot{m}$ or with Q as the dependent variable, the equivalent solutions would have been

$$\dot{m} = k \cdot \frac{\rho r^4}{\mu} \cdot \frac{dp}{dx} \tag{2}$$

and $\quad Q = k \cdot \frac{r^4}{\mu} \cdot \frac{dp}{dx}$ $\hspace{4cm}$ (3)

Suppose now that we attempt to force the issue and work in the three independent length dimensions X, R and Ψ, where R represents the dimension of radial length and Ψ that of tangential length. The corresponding matrix will then be

	M	X	R	Ψ	T
U	0	1	0	0	−1
dp/dx	1	0	−1	−1	−2
ρ	1	−1	−1	−1	0
μ	1	−1	1	−1	−1
r	0	0	1	0	0

Here a singularity develops ($M \equiv - \Psi$) and we obtain the same complete solution as set out in equation 1. However, the position changes as soon as we consider the mass flow rate rather than the mean velocity. In this case no singularity occurs, $(n - r) = 0$ and no DP is produced, leaving the problem, as stated, apparently insoluble.

The reason for this difficulty lies not in the (valid) assumption that R and Ψ are physically independent but in the implicit use of the geometrical relationship $A = \pi r^2$ which is involved in the determination of $\dot{m}$ and which is heterogeneous in $R\Psi$ (cf. the examples of 5.5). This seeming impasse may, however, be overcome by working with two independent characteristic quantities, say the cross-sectional area A and r, a procedure which enables us to take advantage of the independence of R and Ψ. With m as the dependent variable, the resulting matrix will now be

	M	X	R	Ψ	T
$\dot{m}$	1	0	0	0	−1
dp/dx	1	0	−1	−1	−2
ρ	1	−1	−1	−1	0
μ	1	−1	1	−1	−1
A	0	0	1	1	0
r	0	0	1	0	0

The problem is once more capable of solution and we have

$$\dot{m} = k \cdot \frac{\rho r^2 A}{\mu} \cdot \frac{dp}{dx} \qquad (4)$$

At this stage we may, if we so wish, make the numerical (but dimensionally heterogeneous) substitution $A = \pi r^2$ thus reproducing equation 2, but clearly this final result will no longer be homogeneous in $MXR\Psi T$.

It is now clear why, working with $XR\Psi$ and without A as an independent variable, we were able to obtain a complete solution for U but not for $\dot{m}$. This arose from the fact that the value of U is not based directly upon A, whereas $\dot{m}$, being equal to $AU\rho$, is so based.

Our interpretation of the situation is that the two variables A and r may here be considered physically independent provided that no resort, either implied or direct, is made to the equation $A = \pi r^2$. It should also be remarked that equation 4 is more informative than were the previous results in that it reveals the individual and independent contributions made by both A and r to the rate of flow.

We may turn this technique to further advantage by considering the case of laminar flow through an elliptical tube. If we choose the circumference c and the area A as the characteristic variables of the cross-section, we obtain

$$\dot{m} = k \cdot \frac{\rho A^3}{\mu c^2} \cdot \frac{\mathrm{d}p}{\mathrm{d}x}$$

The relevant (heterogeneous) geometrical relationships may now be used to put this result into a more conventional form. With a and b as the major and minor axes of the ellipse, simple substitution gives the complete solution as

$$\dot{m} = k \cdot \frac{\rho a^2 b^2}{\mu} \cdot \frac{\mathrm{d}p}{\mathrm{d}x}$$

This result may be confirmed by conventional analysis. It also holds good for tubes of oval and not necessarily elliptical cross-section, the value of the numerical coefficient k being determined by the precise profile considered.

5.7   The *MFLT* system: Poiseuille's problem (*continued*)

We pointed out in **5.1** that in problems of statics and in situations affected neither by acceleration nor by gravitational attraction, the quantity 'force' is physically independent of mass and it is, in consequence, legitimate to work in the extended *MFLT* system. We illustrate this by reworking the problem considered in the previous section and, taking the mass flow rate $\dot{m}$ as the dependent variable, we have the matrix

	M	F	L	T
$\dot{m}$	1	0	0	−1
$\mathrm{d}p/\mathrm{d}x$	0	1	−3	0
ρ	1	0	−3	0
μ	0	1	−2	1
r	0	0	1	0

Here we observe that both $\dot{m}$ and ρ, relate to 'quantity of matter' and have their dimensions based upon M. It will, moreover, be clear that, since in this situation we are considering only unaccelerated motion, there is no physical dependence of these two quantities upon the applied forces.

The quantities dp/dx and μ, however, are derived from the concept of force. They are physically dependent upon one another in that the pressure gradient is wholly absorbed in overcoming the resistance due to the viscosity of the fluid – and not in accelerating its mass. The dimensions of dp/dx and μ may, therefore, be based on F. Inspection now shows that there is 1 DP and this leads to the complete solution already obtained in equation 2 of **5.6**.

We re-emphasise that an essential condition for the use of *MFLT* is that the motion of the fluid is unaccelerated. Should we require, for example, to examine the drag due to a cylindrical obstacle in a stream, then the dimension F will no longer be physically independent of M in view of the fact that the stream lines are not parallel. Acceleration of the fluid will take place with the concomitant onset of inertial forces which rule out the use of M and F as independent reference dimensions.

5.8   Disparate forms of energy

As pointed out in **5.2**, the set *MLT* may be extended to comprise the dimension of any form of energy, provided the situation be such that the form of energy considered is physically independent of ML^2T^{-2}, the dimensions of mechanical energy. This state of affairs frequently occurs in the fields of heat and electricity. Indeed, we have already considered a thermal quantity, temperature θ, as a reference dimension in **5.3.5**, and no difficulty was encountered because the matrix confirmed our suspicion that θ was in fact independent of the other dimensions which were used.

We now content ourselves with two straightforward, if trite, examples involving the quantity 'heat', preferring to postpone to Chapters 6 and 8 the consideration of more detailed cases until we have dealt with certain subtleties that arise in the treatment of thermal and electrical quantities.

Asking what is the gain in heat (thermal energy), Δh, that develops in a body of mass m that falls to the ground from a height s without bouncing, we draw up the matrix

	M	L	T
Δh	1	2	-2
s	0	1	0
g	0	1	-2
m	1	0	0

which yields $\Delta h = k.mgs$

In this situation, mechanical energy is directly transformed into heat and accordingly we represent the dimensions of heat as ML^2T^{-2}. Had we attempted to represent heat as an independent dimension in its own right, H, we should have landed in an impasse because H would then have been formally independent of MLT whereas the situation clearly entails dependence. The only way out of such a situation would then have been to introduce into the list of variables an *ad hoc* dimensional constant $J \equiv M^{-1} L^{-2} T^2 H$ in order to act as a transformation co-efficient between the two forms of energy which we had decided to treat as formally independent. J is, of course, the 'mechanical equivalent of heat' and its introduction leads to no increase in the power of the analysis.

If, however, the situation is such that heat is conserved without transformation, then we can and should treat H as an independent reference dimension. To illustrate this, we determine the quantity of heat q developed in time t by a gas ring into which gas of density ρ and calorific value c is flowing at a volume flow rate of Q. The matrix here is

	M	L	T	H
q	0	0	0	1
t	0	0	1	0
ρ	1	-3	0	0
c	-1	0	0	1
Q	0	3	-1	0

giving one DP only and

$$q = k \, . \, c\rho Qt$$

Had we worked with this particular set of variables and failed to take advantage of the physical independence of H, we should have been tempted to rely upon the set MLT which leads to 2 DPs and an incomplete solution.

5.9   Conclusion

The general principles involving the possibilities of a useful extension of the number of members of the set of reference dimensions will now be clear. We emphasise, however, that these possibilities are not limited to those mentioned in the foregoing text and that there is considerable scope for ingenuity when dealing with individual problems.

Thus there is no need to use electrical energy as the extended dimension in electrical situations. Provided that the conditions of physical independence are satisfied, we may work with charge, current, potential, permeability or any similar quantity as the additional member of our reference set. And, as we have already seen, it may well be convenient in thermal problems to use the dimension of temperature Θ rather than that of heat H. Indeed, under certain conditions Θ may even be physically independent of H, and it will then be possible to use the two additional reference dimensions and operate successfully with the system $MLTH\Theta$ (6.3.3).

Again, where the situation warrants, there is no reason why we should not work with two physically independent time dimensions T_1 and T_2. As a simple example we observe that the concept of power essentially involves two aspects of time, the time associated with the quantity work $\equiv ML^2 T_1^{-2}$, and the time associated with the duration over which that work is performed $\equiv T_2$. This gives the dimensions of power in the MLT_1T_2 system as $ML^2 T_1^{-2} T_2^{-1}$. If now we seek to determine the power P required to lift a mass m to a height h in time t, we have

	M	L	T_1	T_2
P	1	2	−2	−1
m	1	0	0	0
h	0	1	0	0
g	0	1	−2	0
t	0	0	0	1

Note that the dimension of time entering into the quantity g will be T_1 since this will be physically independent of T_2 which corresponds with the duration of the process. The matrix results in the solution

$P = k \cdot mgh/t$

and we see that working in MLT with the 5 quantities considered would have resulted in 2 DPs and an incomplete solution.

It is, however, not easy to find particularly significant examples of this approach because, if two processes take place in accordance with physically independent time scales, there is a tendency for the situation to degenerate into two disparate and unconnected problems.

The possibilities of extensions to the reference set should, then, be born in mind even when dealing with purely mechanical situations and it is often advantageous to devise an *ad hoc* extension to suit a particular case. Thus we could have worked the problem of **5.3.1**, dealing with the application of a couple to a prism, by resort to the system $MXYZT\alpha$ where α denotes the reference quantity 'angle'. Defining strain in terms of the angle through which the prism is twisted and noting that this was physically independent of all quantities based solely in terms of $MXYZT$, we should have arrived at a complete solution.

6

Dimensions of Thermal Quantities

6.1 The dimensions of heat and temperature

The first law of thermodynamics states that heat and work are equivalent; and Clerk Maxwell in his *Theory of Heat* pointed out that units of heat, such as the calorie, are superfluous, for in no other branch of physics are special units of energy employed. This superfluity of special units was generally admitted when, in 1948, the SI units were introduced; 'quantity of heat' is consequently now measured directly in joules. It follows that it is no longer necessary to work with a 'mechanical equivalent of heat' relating, say, ergs to calories.

We define the quantity of heat contained in a body as the kinetic energy of translation, rotation or vibration of the molecules or atoms which that body comprises; and we say, therefore, that the dimensions of heat are the same as those of work or energy, that is ML^2T^{-2}. Nevertheless, despite this formal equivalence, there are many situations in which heat does not undergo conversion into other forms of energy and, under such conditions, we may, as was shown in Chapter 5, treat heat as an independent reference quantity having the dimension H, thus procuring the advantages of an extension of the dimensional system. All this is straightforward enough. There is, however, liable to be confusion in determining the dimension of temperature Θ in terms of MLT. Two approaches are usually adopted, each being based upon the kinetic theory of gases:

1. The absolute temperature of a gas may be defined as equal to the mean kinetic energy of the molecules, that is the total energy divided by the number of molecules, possibly multiplied by a dimensionless constant designed to procure a convenient size for the unit degree. It follows from this definition that $\Theta \equiv ML^2T^{-2}$.

2. Alternatively, the absolute temperature of a gas may be defined as equal to the kinetic energy of the molecules per unit mass of gas, again possibly multiplied by a suitable constant. This second definition would imply that $\Theta \equiv L^2T^{-2}$, and that temperature is proportional to the mean-square molecular velocity.

These two results are clearly contradictory and we have to determine whether temperature may more satisfactorily and consistently be defined by dividing a mean kinetic energy by a *number* of molecules or by a *mass* of molecules.

Our intuition may suggest that, in the MLT system, temperature and heat should not be regarded as equidimensional. We tend to regard the quantity of heat contained within a body as proportional to its temperature multiplied by its mass and to say that, at any fixed temperature, 2 kg of a substance contains 'twice as much heat' as does 1 kg of the same substance. This suggests that heat and temperature are dimensionally related by the equivalence $H \equiv M\Theta$, which gives the dimensions of temperature as $L^2 T^{-2}$.

There is, however, a crucial objection to this. Consider a vessel containing a mixture of two gases in thermal equilibrium. Let the molecules of one gas be heavy while those of the other are light. Now the whole mass of the gaseous mixture will be at the same temperature and the principle of equipartition of energy suggests that this state of affairs will be reflected by the equality of the mean kinetic energies of each type of molecule. In particular, the uniformity of temperature will certainly not be associated with any equality of mean molecular velocities of translation, for the molecules of the heavier gas will be moving more slowly than are those of the lighter.

In view of this observation we have no hesitation in preferring to regard temperature as having the dimensions of energy rather than of (velocity)^2 and, indeed, this fits in neatly with the elementary kinetic theory of gases where we have the equations

$$pV = \tfrac{2}{3}n\overline{E} = \tfrac{2}{3} \cdot (\tfrac{1}{2} mu^2) = R\Theta$$

Here n is the number of molecules, each of mass m, in a volume V, $\overline{E}$ is the mean molecular kinetic energy and u^2 is the mean-square velocity.

It seems then that, whereas it was suggested that the 'quantity of heat' contained in a body is proportional to the mass of that body, we would now prefer to say that it is proportional to the number of molecules which that body comprises. The approach which we have decided to discard is not in any way 'wrong' and it can be developed without inconsistency. As so often in dimensional analysis, the criteria which we adopt are partly aesthetic and partly heuristic.

Some authors, such as Duncanson[17], have argued that while temperature may certainly be proportional to kinetic energy, it is not necessarily equal to it, and that we do not know that the constant of proportionality is dimensionless. We see no grounds for this difficulty for, just as we have defined heat as a form of energy, so now do we find it convenient to define temperature in similar terms. This is as much a defining relationship as is the equation $f = ma$, and it follows that there is no need for the introduction of any dimensional constant.

The magnitude of the numerical units of a scale of temperature certainly may, and does, differ from that of the units of kinetic energy, but this is merely a matter of convenience in the selection of a scale and has nothing to do with the physics of the situation. In order to transform units of molecular energy increase into, say, kelvin, we need only a numerical conversion factor, as when we transform inches into centimetres. In general, the principle of Occam's Razor applies and we are reluctant to introduce any dimensional constant unless it be required to resolve an inconsistency.

6.2   The $MLT\Theta$ and other systems

It follows from these remarks that, where no transformation between heat and mechanical energy takes place, we may profitably extend the number of reference quantities in our set by at least one. The only criterion is that the new quantity must itself be defined in terms of heat or temperature. The most natural approach will clearly be to use as that quantity either heat itself, involving the $MLTH$ system, or temperature, involving the $MLT\Theta$ system. Although there is little to choose between these, current practice favours the use of $MLT\Theta$ and most modern authors tend to work with this unless there be special reasons for adopting an alternative procedure.

We are now free to deduce the dimensions of the various thermal quantities and, working from the defining equations, no particular difficulties are encountered except, possibly, in the case of the gas constant. Here we write the equation of state of the ideal gas as $pv = R\theta$, where v is the specific volume, that is $v = V/m = 1/\rho$. The dimension of the gas constant is accordingly given as $R \equiv L^2 T^{-2} \Theta^{-1}$. The dimensions of $\mathfrak{R}$, the universal (molar) gas constant are identical with those of R and may be derived from the equation $\mathfrak{R} = MR$, where M is the molecular weight ($\equiv [1]$) of the gas considered.

Note that in SI units R is expressed as kJ/kg K, whereas $\mathfrak{R}$ is expressed as kJ/kmol K. These differ by a factor of kmol/kg, which is, of course, dimensionless. Note further that in the MLT system both R and $\mathfrak{R}$ will be of dimensions M^{-1} as follows from the substitution of $\Theta \equiv ML^2 T^{-2}$ in the above argument.

The dimensions of some other quantities in the $MLT\Theta$ system are:

Entropy: $s \equiv ML^2 T^{-2} \Theta^{-1}$, as follows from the defining equation $ds = dq_r/\theta$, where q_r represents units of 'reversible' heat.

Internal energy: $U \equiv ML^2 T^{-2}$, as follows from the equation representing the first law of thermodynamics.

Enthalpy: $H \equiv ML^2 T^{-2}$, as follows from the equation $H = U + pV$.

All these quantities are often considered in terms of unit mass and it follows that the dimensions of the corresponding 'specific' quantities will be as shown, but with the exponent of M reduced to zero.

Specific heat: $c \equiv L^2 T^{-2} \Theta^{-1}$, as follows from the defining equation $dq/d\theta = mc$.

Thermal conductivity: $\kappa \equiv MLT^{-3}\Theta^{-1}$, as follows from the defining equation: $dq/dt = A.d\theta/dx$ where dq/dt is the rate of heat flow across an element of cross-sectional area A and due to a temperature gradient $d\theta/dx$.

The representation of the various quantities considered will, of course, be dependent upon the dimensional system used. The dimensions of κ, for example, shown above in the $MLT\Theta$ system, may also be expressed as:

$$
\begin{aligned}
\kappa \;\; &\equiv LT^{-3} &&\text{in the } MLT \text{ system,} \\
&\equiv M^{-1}L^{-3}TH &&\text{in the } MLTH \text{ system,} \\
&\equiv L^{-1}T^{-1}H\Theta^{-1} &&\text{in the } MLTH\Theta \text{ system.}
\end{aligned}
$$

These results and others will be found tabulated for ease of reference on page 204.

We make two comments. Firstly, in determining the dimensional representations shown both here and in the appendix we have adopted a somewhat arbitrary procedure. Working in the $MLTH$ system, any quantity involving H is expressed in terms of H, but where that quantity, or any other quantity, includes Θ in its definition, we replace Θ by $ML^2 T^{-2}$. Conversely, in the $MLT\Theta$ system, any quantities involving H are expressed in terms of $ML^2 T^{-2}$, whereas Θ is retained provided that the unit of temperature occurs in the relevant definition. This results in a consistent approach and avoids the possibility of alternative representations of a quantity which would otherwise arise in view of the mutual dimensional interrelationships existing between MLT, H and Θ.

Secondly it will be noted that both heat and temperature have been retained as reference dimensions in the $MLTH\Theta$ system. The use of this again leads to no inconsistency and may often be justified in view of the fact that there are a number of situations in which there arise quantities which are explicitly defined in units that differentiate between H and Θ. We shall, as a matter of convenience, tend to observe this differentiation, but, in view of the fact that H and Θ are generally physically dependent upon one another, the doubly extended system is liable to lead to a singularity in the matrix and no increase in the sharpness of the analysis necessarily results. It may happen, however, as in **6.3.3**, that conditions occur such that H and Θ may be treated as physically independent. Where this proves to be the case, then clearly a gain in information derives from working in $MLTH\Theta$.

There is a further approach which may occasionally be used to advantage. In certain problems involving thermal quantities it becomes useful to distinguish between the inertial and quantitative aspects of mass, that is between M_i and M_q, as in 5.2. Thus in the $MLTH\Theta$ system we may write specific heat as $c \equiv M^{-1}H\Theta^{-1}$, while in the $M_iM_qLT\Theta$ system we have $c \equiv M_iM_q^{-1}L^2 T^{-2}\Theta^{-1}$. Both these forms are equally useful in that each involves a representation in terms of two dimensions in addition to those of the basic triad MLT. (In justification of the latter representation, note that $H \equiv M_i L^2 T^{-2}$, since it is the inertial aspect of mass which is significant, whereas the unit mass of material which enters into the definition of c is related to quantity of matter and has the dimension M_q.) For the use of this approach see **6.4.1**.

6.3 Some examples involving heat transfer

We now illustrate the use of thermal dimensions in a number of examples.

6.3.1 'Boussinesq's problem'

Boussinesq's problem seeks to determine the heat loss of a body immersed in a fluid flowing past it at a constant velocity. Working with $MLTH\Theta$ and assuming that the fluid is incompressible, the relevant quantities will be as shown in the following table:

Physical quantity	Symbol	M	L	T	H	Θ
Heat-transfer coefficient	h	0	−2	−1	1	−1
Velocity of fluid	U	0	1	−1	0	0
Density of fluid	ρ	1	−3	0	0	0
Specific heat of fluid (c_p)	c	−1	0	0	1	−1
Dynamic viscosity of fluid	μ	1	−1	−1	0	0
Thermal conductivity of fluid	κ	0	−1	−1	1	−1
Linear dimension of body	d	0	1	0	0	0

Commenting upon this, we mention that the heat-transfer coefficient h is defined as the heat transferred by the body per unit area per unit time per unit temperature difference between the body surface and the surrounding fluid. We also note that, as might have been anticipated, the H and the Θ columns are linearly dependent, the one being equal to −1 times the other. There will, therefore, be a singularity in the matrix contained in the table and no advantage has derived from using $MLTH\Theta$ in preference to $MLT\Theta$, other than a certain convenience in the determination of the dimensional representations.

The routine approach yields $(7 - 4) = 3$ DPs as shown in the equation

$$\frac{hd}{\kappa} = \phi \left(\frac{\rho U d}{\mu}, \frac{\mu c}{\kappa} \right) \tag{1}$$

This solution may be made more informative if various specialised situations are considered. Thus, if viscosity is negligible, the two DPs contained in the argument of the function may clearly be combined multiplicatively to yield

$$\frac{hd}{\kappa} = \phi\left(\frac{\rho U d c}{\kappa}\right)$$

A second specialisation may be derived as follows. We shall see in **6.5** that the DP $(\mu c/\kappa)$ which occurs in equation 1 is known as the Prandtl number. Since all the variables contained within it are material properties of the fluid considered, we are frequently able to regard this as constant for a given fluid, in which case we obtain the relationship

$$\frac{hd}{\kappa} = \phi\left(\frac{\rho U d}{\mu}\right)$$

For bodies of various geometrical shapes, empirical solutions have been obtained for the various functions listed, these solutions usually taking the form of a PP.

For situations in which the geometry of the body warrants it, and where either the viscosity is negligible or the flow is laminar, there will be physical independence between orthogonal length dimensions, as may be verified by an examination of the underlying equations (7.2). Even at this stage, however, if we be permitted to abandon the completely rigorous approach, we may obtain further specialised results which are of interest. Let us, for example, consider laminar flow past a flat place aligned with the stream. Working with $MXYZTH\Theta$, the indicial matrix becomes

	M	X	Y	Z	T	H	Θ
h	0	−1	0	−1	−1	1	−1
U	0	1	0	0	−1	0	0
ρ	1	−1	−1	−1	0	0	0
c	−1	0	0	0	0	1	−1
μ	1	−1	1	−1	−1	0	0
κ	0	−1	1	−1	−1	1	−1
d	0	1	0	0	0	0	0

Here, in choosing the exponents of the X, Y and Z dimensions, we have specified the x direction as lying along the line of flow and the y direction as orthogonal to the plane of the plate. It follows from this that the area occurring in the definition of the quantity h will have the dimensions XZ. In determining the representation of κ we take it that the heat flow will be in the y direction; that this is so follows both from our physical intuition and from a consideration of the underlying theoretical equations, which will be homogeneous in XYZ only provided that the heat flow does indeed take place in a direction orthogonal to the plate.

A routine exercise now gives

$$\frac{h^2 d\mu}{\kappa^2 \rho U} = \phi\left(\frac{\mu c}{\kappa}\right) \tag{2}$$

For fully developed flow through a circular tube, the matrix shown above will be modified to show $d \equiv Y$ and analysis in this case yields simply

$$\frac{hd}{\kappa} = \phi\left(\frac{\mu c}{\kappa}\right) \tag{3}$$

Finally, provided we are able to consider the fluid such that the Prandtl number is constant, we can obtain a complete solution for each of the two cases considered, these being

$$\frac{h^2 d\mu}{\kappa^2 \rho U} = \text{constant for a flat plate} \tag{4}$$

and $\quad \dfrac{hd}{\kappa} = \text{constant for a circular tube} \tag{5}$

6.3.2 'Natural' convection from a vertical plate

Consider next the heat emitted by a vertical plate under conditions involving free or 'natural' convection. The indicial matrix will be

	M	X	Y	Z	T	H	Θ
h	0	-1	0	-1	-1	1	-1
ρ	1	-1	-1	-1	0	0	0
μ	1	-1	1	-1	-1	0	0
κ	0	-1	1	-1	-1	1	-1
c	-1	0	0	0	0	1	-1
$\Delta\theta$	0	0	0	0	0	0	1
$g\beta$	0	1	0	0	-2	0	-1
x	0	1	0	0	0	0	0

Here the orthogonal directions considered are: x measured vertically along the plate, y the direction normal to the plate and z horizontally in the plane of the plate. The extended set of dimensions involving XYZ is again to be justified if and only if the boundary layer is laminar (7.2). Other comments made in the case of 6.3.1 continue to apply with the exception of our decision to include the compound variable $g\beta$ in place of U. Our thinking here is that the phenomenon of free convection is concerned with the buoyancy force upon an element caused by thermal expansion, and the characterising quantity must then be $g\beta$, where β is the coefficient of volume expansion defined by $\beta = -(1/\rho)\,(\partial\rho/\partial\theta)_p$. The quantity c, as in the previous example, is the specific heat at constant pressure, c_p.

The rank of the matrix is 6 and there will, therefore, be two DPs only, these appearing in the equation

$$\frac{h^4 x \mu^2}{\kappa^4 g\beta\rho^2 \Delta\theta} = \phi\left(\frac{\mu c}{\kappa}\right) \tag{1}$$

Had the analysis been carried out in terms of the single length dimension L, we would have obtained the less informative solution

$$\frac{hx}{\kappa} = \phi\left(\frac{g\beta x^3 \rho^2 \Delta\theta}{\mu^2}, \frac{\mu c}{\kappa}\right) \tag{2}$$

This latter solution may be applied to the case of turbulent flow, since it does not result from the assumption of physical independence between orthogonal length dimensions.

6.3.3 Periodic temperature changes beneath the earth's surface

We now investigate the manner in which diurnal (or annual) temperature changes penetrate into the rock or soil beneath the earth's surface. Assume that surface temperature may be expressed as a mean temperature coupled with a periodical fluctuation $\theta_a\,\phi(\omega t)$, where ϕ is not necessarily sinusoidal. There will be no loss of generality if we chose a temperature scale such that the datum (zero) is equal to the mean surface temperature. At depth x and at time t the rock temperature θ will then be a function of the relevant material properties of the rock — that is c, κ and ρ — together with the suface fluctuation frequency fully characterised by ω and θ_a.

For illustrative purposes we work in each of the two systems MLT and $MLTH\Theta$, the relevant matrices being

	M	L	T	M	L	T	H	Θ
θ	1	2	-2	0	0	0	0	1
θ_a	1	2	-2	0	0	0	0	1
x	0	1	0	0	1	0	0	0
t	0	0	1	0	0	1	0	0
ω	0	0	-1	0	0	-1	0	0
κ	0	-1	-1	0	-1	-1	1	-1
c	-1	0	0	-1	0	0	1	-1
ρ	1	-3	0	1	-3	0	0	0

In the MLT system, with the 8 quantities considered, we have, as expected, 5 DPs, as shown in the equation

$$\frac{\theta}{\theta_a} = \phi\left(\omega t, \frac{\theta_a}{\rho x^5 \omega^2}, x^3 \rho c, \frac{\kappa x}{\omega}\right)$$

In the extended $MLTH\Theta$ system, however, we notice that the rank of the indicial matrix increases from 3 to 5 and we see, in particular, that in the situation being considered there is no physical dependence between H and Θ. These quantities, may, therefore, be used as independent reference dimensions and no singularity develops. In consequence there are now 3 DPs only and

$$\frac{\theta}{\theta_a} = \phi\left(\omega t, \frac{\rho c \omega x^2}{\kappa}\right)$$

This analysis may be taken further by investigating rather more specific results. Suppose that we are interested in, say, the amplitude of the temperature fluctuations at a constant depth; then, as this will be independent of time, we may omit t from our list of variables and obtain simply

$$\frac{A}{\theta_a} = \phi\left(\frac{\rho c \omega x^2}{\kappa}\right)$$

where A is the required temperature amplitude at depth x. Defining a wavelength λ as the vertical distance between successive temperature maxima (or minima), we may put

$$\lambda = \phi\left(\theta_a, \omega, \kappa, c, \rho\right)$$

and inspection gives a complete solution:

$$\frac{\lambda^2 \rho c \omega}{\kappa} = \text{constant}$$

Similarly, we may define the wave velocity as the velocity at which a fluctuation maximum (or minimum) penetrates into the rock, and once again a simple equation may be determined by the methods indicated above and by making use of the equation $v = \omega\lambda$.

The conventional approach to this problem depends upon the integration of a partial differential equation. (See Joos[36].) This may be written

$$\frac{\partial \theta}{\partial t} = \frac{\kappa}{\rho c} \cdot \frac{\partial^2 \theta}{\partial x^2}$$

Solving, and substituting values for the material properties which are applicable to 'typical' soils and loosely consolidated surface rocks, we find that the depth at which the annual temperature variation is first out of phase is about 4m, that is to say that at this depth the temperature maximises in midwinter and minimises at midsummer. The amplitude of the fluctuations will here be about one sixteenth of the surface amplitude. For diurnal variations, the equivalent depth is of the order of only 0.2 m, with much the same factor for the amplitude decrease.

6.4   Some examples involving kinetic theory

6.4.1   Thermal conductivity of a gas

The derivation of an expression for the thermal conductivity of a gas in terms of its molecular properties is a problem involving as many as 7 variables, yet we will show that a complete solution is possible. The example is based on one given by Huntley[34] and, for illustrative purposes, we work with M_i and M_q — the approach mentioned in 6.2. In preparing the matrix we take X as the length dimension in the direction of heat flow, while L is the length dimension in any direction orthogonal to this. The significant variables are as shown in the table:

Physical quantity	Symbol	M_i	M_q	X	L	T	Θ
Thermal conductivity	κ	1	0	5/3	−2/3	−3	−1
No. of molecules per unit volume	N	0	0	−1	−2	0	0
Molecular mass	m	0	1	0	0	0	0
Mean molecular velocity	v	0	0	1	0	−1	0
Mean free path	λ	0	0	1	0	0	0
Gas pressure	p	1	0	−1/3	−2/3	−2	0
Specific heat of gas	c_V	1	−1	2/3	4/3	−2	−1

Commenting upon this, it will be noticed that we have concerned ourselves with energy transfer in the x direction and it follows that the dimensions of v and of λ will involve X only. The specific heat at constant volume; c_V, is included among the molecular properties as it is a measure of the increased energy of the molecules due to heating, and the dimensions shown for this quantity correspond with those given in 6.2. Similarly the dimensions of κ are deduced from the representation previously given, that is $\kappa \equiv XL^{-2}T^{-1}H\Theta^{-1}$.

This latter formula clearly corresponds to the case of heat transfer in the x direction. In order that this may be adapted to the $M_i M_q X L T \Theta$ system, we introduce the substitution $M_i X^{2/3} L^{4/3} T^{-2}$ for H. (Here the length dimensions $X^{2/3} L^{4/3}$ are in accord with the principle of symmetry and correspond to an equal distribution in each of the three orthogonal directions.)

We now have $(7 - 6) = 1$ DP and our solution is:

$$\kappa = k \,.\, mNv\lambda c_V$$

with the quantity p dropping out since its exponent is zero. Note further than (mN) is equal to the gas density and that the absolute temperature is, by definition, proportional to v^2. Taking these two relationships into consideration, the equation may be rewritten in its final form

$$\kappa = k \,.\, \rho\lambda c_V \theta^{1/2}$$

This is a good approximation to the actual position, although, in fact, the thermal conductivity of a gas is found to increase rather more rapidly than does the square root of the absolute temperature. Observations on neon, for example, show that $\kappa \propto \theta^{0.7}$ over the range $-181°$C to $106°$C. This departure from the square-root law derives from the fact that the molecules should properly be regarded, as in our next example, as centres of repulsive force rather than as the 'hard spheres' of elementary kinetic theory.

We also point out that the equation is not applicable at very low pressures, for under such conditions κ is found to be pressure-dependent. In commenting upon this difficulty, we would say that at low pressures a different selection of significant quantities has to be made, for the mean free path is then long compared with the dimensions of the containing vessel and λ therefore no longer affects the situation.

This observation emphasises, yet again, the need for care and insight in making an appropriate selection of variables. It must be remembered that dimensional analysis can only say, '*If* these quantities enter into a relationship, *then* that relationship is necessarily of such and such a nature.' Dimensional analysis can never say, 'It is these particular quantities, and these alone, which are of significance in this or that special situation.' In order to determine whether or not a quantity is in fact relevant, we have to draw upon our experience and upon a sometimes considerable maturity of physical insight.

6.4.2   *Repulsive force between molecules in a gas*

An example of a different type illustrates the manner in which dimensional analysis may be used as a tool for the interpretation of experimental results. This relates to the classical investigation of Rayleigh[57] into the effect of temperature on a gas and into the associated intermolecular forces.

Simple kinetic theory, based on the assumption that molecules behave as small and perfectly elastic spheres, gives the result

$$\mu = \tfrac{1}{3}\, mNv\lambda = \tfrac{1}{3}\, \rho v\lambda$$

v being the mean molecular velocity. Since λ is inversely proportional to ρ, this suggests that viscosity is independent of both quantities and that this is so has, indeed, been confirmed observationally within the pressure range of, say, 0.02 to 1.0 atmospheres.

At a more sophisticated level we discard the 'hard-sphere' hypothesis and assume that the molecules, instead of having a definite size, behave as centres of a strong repulsive force when in close proximity to one another. We further assume that forces of attraction are negligible and that the repulsive force obeys an inverse power law. Since, as we have already seen, μ will be largely independent of λ and ρ, we may expect it to be dependent upon v, m and a repulsion coefficient K defined by the equation

$$f = K \cdot r^{-n}$$

Here f is the repulsive force between two molecules separated by a distance r.

It follows that K will be a dimensional coefficient characteristic of the molecules under consideration and n will be the unknown exponent to be determined. We therefore prepare the indicial matrix

	M	L	T
μ	1	-1	-1
m	1	0	0
v	0	1	-1
K	1	$n+1$	-2

Here the dimensions of the repulsion coefficient, being taken from the equation $f = K \cdot r^{-n}$, are dependent upon the unknown exponent n which it is our task to determine. There is one DP only and we have

$$\mu^{n-1} = k \cdot m^{n+1} \, v^{n+3} K^{-2}$$

Taking temperature to be equal to the mean kinetic energy of the molecules, we make use of the relationship $\theta = \frac{1}{2} m v^2$ to eliminate v, obtaining:

$$\mu^{n-1} = k \cdot m^{(n-1)/2} K^{-2} \, \theta^{(n+3)/2}$$

Since for a given gas m and K will be constant, it follows that the viscosity–temperature relationship will be

$$\mu = k \cdot \theta^{(n+3)/2(n-1)}$$

This shows that μ is proportional to some power of θ. All that remains, then, is to determine experimentally the relationship between these two variables. Subsequently, when we plot the results on logarithmic paper, the gradient of the line obtained will indicate the required power of θ. To quote two cases, it is found that for argon $\mu = k \cdot \theta^{0.815}$ while for hydrogen $\mu = k \cdot \theta^{0.681}$. The equivalent values for n, the exponent occurring in the original equation $f = K \cdot r^{-n}$, will then be 7.35 (for argon) and 12.05 (for hydrogen).

As a final comment we may mention that for the case of a 'hard-sphere' molecule, the value of the exponent n becomes effectively infinite and we notice that this leads to a viscosity–temperature relationship $\mu = k \cdot \theta^{1/2}$, which is, indeed, in accordance with the equation deduced from conventional analysis based on elementary kinetic theory.

6.4.3   The critical temperature of He³

A not dissimilar approach involving part dimensional analysis and part observation was carried out by de Boer[2] in order to determine the critical temperature of the rare isotope of helium, He³.

Recall that the equation of state for the N molecules of a perfect gas may be written in the form $pv = k\theta$, where v here is the mean volume occupied per molecule, that is V/N, and k is Boltzmann's constant.

We now assume, rather more realistically than we did in our previous example, that, while there is a strong repulsive force between molecules in the immediate vicinity of one another, there is also a weak attractive force between them at larger distances. Following de Boer in essentials, we argue that the potential energy deriving from this force may be represented by the curve in figure 9, and we are

Figure 9

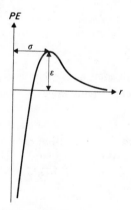

able to characterise the situation by the introduction of two parameters: ϵ, being the maximum potential energy deriving from the repulsive force that develops and σ, being the distance between molecules at which this maximum takes place.

This leads us to a more general form of the equation of state which we write:

$$f(p, v, k, \theta, \epsilon, \sigma, m) = 0$$

where m, as before, denotes the molecular mass. Working with $MLT\Theta$ this equation leads to 3 DPs, namely

$$P^* = (\sigma^3 p/\epsilon) \quad V^* = (v/\sigma^3) \quad \Theta^* = (k\theta/\epsilon)$$

and it follows that the 'generalised equation of state' will take the form

$$f(P^*, V^*, \Theta^*) = 0$$

This represents the 'law of corresponding states' and may conveniently be illustrated by a contour diagram as shown in Figure 10. Note in particular the

Figure 10

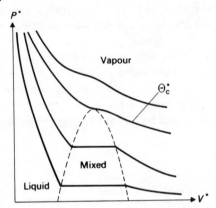

critical isotherm Θ^*_c at and above which the liquid and vapour phases merge with changes in the value of P^* and without the occurrence of any intermediate state involving mixed phases.

At this stage de Boer pointed out that the argument should be based upon a quantum rather than upon a classical interpretation of the situation, since it is to be expected that Planck's constant h enters the situation, particularly where light molecules are involved. He therefore introduced the further DP

$$\Lambda^* = \frac{h}{(\sigma^2 m\epsilon)^{1/2}}$$

which he included in the 'generalised equation of state', rewriting this as

$$f(P^*, V^*, \theta^*, \Lambda^*) = 0$$

An implication of this is that each gas is characterised by a particular value of Λ^* and that the critical temperature θ^* is no longer a constant but is expressible as a function of Λ^*.

In brief outline, de Boer observed the departures from Boyle's law at high temperatures for a number of gases and was thus able to estimate corresponding values of σ and ϵ. This enabled him to make a plot of θ^*_c against Λ^* which could readily be extrapolated to include He^3 with its mass number of 3 (instead of 4 as in the case of the commoner isotope He^4). From this curve, which is shown roughly sketched in figure 11, he was able to read off the expected value of θ^*_c for

Figure 11

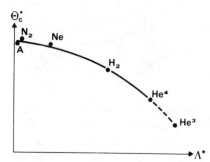

He^3 and this, when converted into the critical temperature, gave $3.3K$ ($\pm 0.2K$). Some years later, when eventually He^3 had been prepared in a nuclear reactor in sufficient quantity for its critical temperature to be measured, the value was found to be precisely in the centre of the range predicted by de Boer. ●

6.5   The use of standard dimensionless numbers

In the analysis of problems involving thermal (and other) quantities, we repeatedly encounter a number of 'standard DPs'. It will be convenient to list a selection of these, since a familiarity with them will often enable us to write down directly the various DPs associated with a set of variables instead of resorting to the possibly laborious analysis of the relevant indicial matrix. It will be understood, however, that while acquaintance with these DPs, or standard dimensionless numbers, will be a useful tool, it is in no case an essential one for the successful analysis of a problem.

With much of the notation already established, we now list the more common numbers appearing in problems involving heat and fluid dynamics:

Table 1   Standard DPs

Name	Symbol	Definition
Eckert number	Ec	$U^2/c_p\Delta\theta$
Froude number	Fr	$U/(gl)^{1/2}$
Grashof number	Gr	$g\beta\Delta\theta l^3\rho^2/\mu^2$
Mach number	M	U/a
Nusselt number	Nu	hl/κ
Prandtl number	Pr	$\mu c_p/\kappa$
Reynolds number	Re	$\rho Ul/\mu$
Stanton number	St	$h/\rho Uc_p$
Strouhal number	S	nl/U
Weber number	Wb	$\rho U^2 l/\tau$

(Here β is a coefficient of volume expansion, a the velocity of sound, n a frequency and h a heat transfer coefficient. The remaining symbols will be familiar.)

As an example of the use of these numbers, we proceed to carry out an analysis of the following situation. Consider a class of geometrically similar water-tube boilers characterised by the tube diameter D, and assume that the heat-transfer coefficient h governing the transfer of heat from the flue gases to the tubes is a function of v, D, μ, κ, ρ and c_p, where v is the velocity of the gases. It follows by inspection of the above table that a complete set of DPs will then consist of those appearing in the argument of the function f(Re, Pr, Nu) = 0, and this completes the essentials of the required analysis. Further investigation, or specialisation, may then be carried out at will.

Again, in Boussinesq's problem (**6.3.1**), equation 1 representing the general case may be written simply as

$$Nu = \phi\,(Re, Pr)$$

Equation 2 representing the case of the laminar boundary layer may be written

$$Nu/Re^{1/2} = \phi\,(Pr)$$

or, under conditions where Pr may be taken as constant

$$Nu = k \cdot Re^{1/2}$$

Similarly, equation 3, which gave an expression for heat transfer from fully developed laminar flow in a circular tube may be written

$$Nu = \phi(Pr)$$

and we see that the Reynolds number does not enter into this particular situation. Finally, under conditions where Pr may be taken as constant, we have equation 5 of the same problem which reduces very simply to

Nu = constant

If a further illustration be required, we may revert to **6.3.2**, where we considered free convection from a vertical plate. We now see that equations 1 and 2 of that problem may be written respectively as

$$Nu/Gr^{1/4} = \phi\,(Pr)$$

and $Nu = \phi\,(Gr,\,Pr)$

The use of standard dimensionless numbers will also be found particularly applicable in problems met with in fluid mechanics, and this approach will, therefore, be considered further in Chapter 7.

There will be found listed in the McGraw-Hill *Encyclopaedia of Science and Technology*[45] some three hundred 'named dimensionless groups'. The use of most of these is restricted to various specialist branches of physics and technology and they will be of but little interest to the general reader.

7

Dimensional Analysis in Fluid Dynamics

7.1 Dimensionless numbers in fluid dynamics

Although the present discussion introduces no new physical quantities or dimensions, it is advisable to devote a separate chapter to fluid dynamics because dimensional analysis is probably used more advantageously in this field than in any other. That this is so is largely due to the intractable nature of the mathematics involved in any more direct approach.

Extensive reliance is placed on standard DPs. Of those mentioned in 6.5 perhaps the most frequently used are Re, Fr, M and Wb. Now it will be recalled from 1.8 that any DP may be regarded as a ratio of similar quantities, and it is useful to consider the DPs of fluid dynamics as ratios of forces. To bring this out, we tabulate the four DPs listed above, using for M and Wb an alternative but equivalent definition to that given in Table 1.

Table 2   The commoner DPs of fluid dynamics

Symbol	Definition	Force ratio
Re	$\rho Ul/\mu$	inertia/viscous
Fr	$U/(gl)^{1/2}$	inertia/gravitational
M	$U/(K/\rho)^{1/2}$	inertia/compressive
Wb	$U/(\tau/\rho l)^{1/2}$	inertia/surface tension

Each of these DPs may, be derived by an exercise in dimensional analysis performed upon the variables involved, but the point we are emphasising here is that it is often helpful to seek a physical interpretation of the significance of the number concerned. Thus the manner in which Re may be regarded as a ratio between the intertial and the viscous forces existing within a fluid will be fully discussed in 10.2.

The reader will notice the persistent inclusion of an inertial term in the force ratios listed in Table 2 and we do, in fact, observe that this term is almost invariably significant in the more commonly occurring situations, whereas only a very restricted selection of other terms are generally applicable. It follows that a DP such as $(d^3 g/\nu^3)$, while admissible, finds no general use.

An alternative approach recognises that it may sometimes be convenient to regard the DPs listed as velocity ratios. As is well known, the Mach number represents the ratio of a characteristic flow velocity U to the velocity of small amplitude compressive waves a ($=(K/\rho)^{1/2}$). Again, the Froude number may be thought of as the ratio of a characteristic velocity to a small-amplitude surface-wave velocity. Interesting though this approach may be, we will not pursue it further as it would represent an excursus from our main argument.

Although a distinction between dependent and independent variables is not always possible, the DPs listed in Table 2 are generally constructed from the independent variables characterising a particular flow. The dependent DP, that is the DP containing the dependent variable, will often be associated with the fluid forces acting on a body. For reasons outlined above, this will also most conveniently be incorporated into a dimensionless group by combining it with an inertial term. Further conventions influencing the selection of 'dependent' DPs may be briefly mentioned:

1. The inertial term is generally taken as $\frac{1}{2}\rho U^2$, an expression which is considered to have important physical significance.

2. The area used in forming force coefficients is orthogonal to the free stream direction. This convention will, for example, eliminate inconsistencies when considering two-dimensional flow.

3. As the absolute pressure on a body is less significant than is the relative pressure, the pressure coefficient is based on $(p - p_s)$ rather than on p alone. (Here p_s is the static pressure in the free stream.)

With these conventions in mind, we construct Table 3 showing the more common 'dependent' DPs:

Table 3   Standard 'dependent' dimensionless numbers used in fluid dynamics

Name	Symbol	Definition
Drag coefficient	C_D	$D/(\frac{1}{2}\rho U^2 A)$
Lift coefficient	C_L	$L/(\frac{1}{2}\rho U^2 A)$
Skin-friction coefficient	C_f	$\tau_w/(\frac{1}{2}\rho U^2)$
Pressure coefficient	C_p	$(p - p_s)/(\frac{1}{2}\rho U^2)$
Moment coefficient	C_M	$M/(\frac{1}{2}\rho U^2 Al)$
Strouhal number	S	nd/U

In this table, D and L represent forces directed respectively along and at right angles to the direction of free stream flow, while τ_w denotes the shear stress resulting from the flow and measured at the boundary (wall) of the submerged body. The Strouhal number has been included to accommodate cases in which the flow is characterised by a periodic motion, which may be either dependent or independent. The frequency n of this motion may then be expressed in some such dimensionless form as (nd/U); in a problem involving, for example, the determination of the eddy shedding frequency behind a bluff body in uniform flow, we find that, since n is dependent upon ρ, U, d and μ, we have simply

$$(nd/U) = \phi(\rho U d/\mu)$$

or $s = \phi(Re)$

Note that the choice of (nd/U) as one of our DPs is a matter of convention only in that this is a standard form. There would, of course, have been no loss of information had another DP such as $(nd^2 \rho/\mu)$ been used.

Again, if we are interested in the drag per unit length of a circular cylinder in a high-speed uniform flow, we have to consider the drag per unit length D' as dependent upon ρ, U, d, μ and K. A variety of DPs may be formed from these variables but, in accordance with the conventions already outlined, we carry out the analysis in terms of standard DPs to obtain

$$\frac{D'}{\frac{1}{2}\rho U^2 d} = \phi \left(\frac{\rho U d}{\mu} , \frac{U}{(K/\rho)^{1/2}} \right)$$

or, very simply

$$C_D = \phi \left(Re, M \right)$$

7.2   Physical independence between orthogonal length dimensions

In certain problems it may be possible to make profitable use of the extended set of length dimensions discussed in Chapter 5. The criteria for physical independence between dimensions were set out in 5.2, but in many situations in fluid dynamics this independence is particularly difficult either to establish or to refute, and it becomes necessary, therefore, to carry out a more detailed study of the conditions under which it becomes possible to work with XYZ rather than merely with L. We consider the question from the mathematical rather than from the physical viewpoint and we base our argument on the fact that a necessary and sufficient condition for the consistent use of XYZ is clearly that the basic equation relating to the situation considered must be homogeneous in the dimensional set adopted. Furthermore, homogeneity must be preserved in any subsequent manipulations that may be applied to the basic equations. (See 5.5 and, particularly, the accompanying examples.)

This approach is, then, applied to the Navier–Stokes and continuity equations which are descriptive of incompressible fluid flow and which may be written:

$$\frac{\partial u}{\partial t} + u\,\frac{\partial u}{\partial x} + v\,\frac{\partial u}{\partial y} + w\,\frac{\partial u}{\partial z} = \frac{1}{\rho}f_x - \frac{1}{\rho}\,\frac{\partial p}{\partial x} + v\left(\frac{\partial^2 u}{\partial x^2} + \frac{\partial^2 u}{\partial y^2} + \frac{\partial^2 u}{\partial z^2}\right)$$

$$\frac{\partial v}{\partial t} + u\,\frac{\partial v}{\partial x} + v\,\frac{\partial v}{\partial y} + w\,\frac{\partial v}{\partial z} = \frac{1}{\rho}f_y - \frac{1}{\rho}\,\frac{\partial p}{\partial y} + v\left(\frac{\partial^2 v}{\partial x^2} + \frac{\partial^2 v}{\partial y^2} + \frac{\partial^2 v}{\partial z^2}\right)$$

$$\frac{\partial w}{\partial t} + u\,\frac{\partial w}{\partial x} + v\,\frac{\partial w}{\partial y} + w\,\frac{\partial w}{\partial z} = \frac{1}{\rho}f_z - \frac{1}{\rho}\,\frac{\partial p}{\partial z} + v\left(\frac{\partial^2 w}{\partial x^2} + \frac{\partial^2 w}{\partial y^2} + \frac{\partial^2 w}{\partial z^2}\right)$$

$$\frac{\partial u}{\partial x} + \frac{\partial v}{\partial y} + \frac{\partial w}{\partial z} = 0$$

Here u, v and w are the fluid velocity components in the x, y and z directions and f_x, f_y and f_z are the components of any body forces which may be operative.

Examination shows that there is no homogeneity in the extended XYZ system since the dimensions assigned, say, to p and v will in general vary from point to point, thus contradicting the requirement of a fixed dimensional representation valid throughout the situation. Alternatively, any attempt to distinguish between orthogonal components will be invalid as isotropy is generally assumed at some stage, with the consequent implication that, say, $p_x = p_y = p_z$. It will then be inevitable that dimensional homogeneity is lacking unless in some special case the pertinent variables are either absent or susceptible to a constant dimensional representation.

In amplification of this remark, we see that in a generalised flow the viscosity may be considered to be 'acting' with different degrees of influence in different directions and it will not, then, be possible to express this quantity with a specific dimensional representation in the XYZ system. If, for example, we consider the problem of determining the drag due to a steady creeping motion past a sphere, an examination of the mathematically derived solution (Schlichting[64]) clearly shows that viscous forces are acting in more than one direction, and there is in consequence no dimensionally homogeneity when the extended set is used.

This situation may be contrasted with that which arises where there is steady laminar flow between parallel plates. In this case pressure and viscosity will each have a unique preferred (axial) direction, as may be seen from both physical and mathematical standpoints. If, then, we take the direction of flow as in the x direction, the Navier–Stokes and continuity equations reduce to

$$-\frac{1}{\rho}\,\frac{dp}{dx} + v\,\frac{d^2 u}{dy^2} = 0$$

Here the kinematic viscosity v has the dimensions $Y^2 T^{-1}$ and homogeneity is maintained. It follows that the extended set $MXYZT$ may here be used to advantage.

From these remarks, it will be seen that although the equations of motion do not in general remain homogeneous in $MXYZT$, there will, nevertheless, be many special cases in which homogeneity does, in fact, persist. It will, then, be instructive to consider a number of individual cases and to determine whether or not resort may legitimately be made to an extended dimensional system. In our discussion we make the assumption that no body forces are acting unless the contrary is specifically stated.

1. Ideal fluids If viscous terms are eliminated from the Navier–Stokes and continuity equations, we obtain equations descriptive of the motion of an 'ideal' fluid. As is well known, such motion is irrotational, a condition which is realised mathematically by the equations

$$\frac{\partial u}{\partial y} - \frac{\partial v}{\partial x} = 0$$

$$\frac{\partial v}{\partial z} - \frac{\partial w}{\partial y} = 0$$

$$\frac{\partial w}{\partial x} - \frac{\partial u}{\partial z} = 0$$

These equations are certainly not homogeneous in XYZ and it follows that the use of $MXYZT$ is inadmissible in situations involving ideal fluids.

We make an important comment. In vector notation the condition of irrotationality is given by $\nabla \times u = 0$, where u is the point velocity vector of the fluid. This relationship is a necessary and sufficient condition for the existence of a scalar potential ϕ such that $u = -\nabla\phi$. (See, for example, Rutherford[61].) If, now we ignore the time-dependent and pressure terms of the Navier–Stokes equations and rewrite them in terms of the scalar potential, we obtain the Laplace equation: $\nabla^2 \phi = 0$. This equation, which again has no homogeneity in XYZ, has wide applications in many branches of physics including, for instance, heat flows and electric fields. We are, therefore, able to state that in none of these cases, many of which are of the greatest importance, can the extended set of length dimensions be resorted to unless simplifications are made or specialisations considered.

2. Compressible fluids The Navier–Stokes equations may be rephrased in such a way that they become descriptive of compressible fluid flow. However, in many situations in which compressible flow is encountered, viscosity is generally regarded as negligible and irrotationality is, in consequence, assumed. It follows that problems involving compressible flow are, once again, not generally susceptible to analysis in terms of XYZ.

3. Turbulent flows If a flow is turbulent there will be fluctuating components superposed upon, say, a uniform flow velocity U taking place in the x direction. The turbulent flow will then have velocity components equal to $(U + u')$, v' and w' in the x, y and z directions respectively.

If dimensional homogeneity in the Navier–Stokes equations is to be maintained, there will be an inferred independence between the fluctuating velocity components themselves, coupled with a dependence between each such component and the corresponding length dimension. This requirement would, for example, imply that if the y direction were scaled up relative to the x direction, then v' would have to be similarly transformed in relation to u'. Moreover, any considerations involving the statistical properties of these component velocities would not alter the position. Turbulent flows, however, are not generally characterised by such behaviour and are, in fact, isotropic. It follows that they cannot be legitimately analysed in terms of extended sets of length dimensions. More directly, the equations which have to be satisfied when isotropic turbulent flow takes place are themselves heterogeneous in XYZ since $u'^2 \neq v'^2 \neq w'^2$, and the method breaks down.

4. Effect of body forces The presence of body forces will not of itself introduce any lack of homogeneity provided that the appropriate length component be used in each term as it arises. A problem involving gravitational forces will be found in Appendix 2 (problem 26) and this may be solved by resort to an extended length system.

5. Heat transfer If thermal effects are significant, the viscosity, which is dependent upon temperature, will vary with x, y, z and t. The Navier–Stokes equations may be rewritten to cover this situation, but a further equation, derived from considerations of energy transfer, will also be required.

Once again, although the generalised equations are themselves heterogeneous in XYZ, a number of particular cases will exhibit homogeneity and the use of extended length dimensions may then be used to advantage. The problem of laminar flow through a heated tube has, for example, already been successfully treated in **6.3.1** and the complete solution obtained would not have been available unless resort had been made to the extended set involving XYZ.

6. Laminar boundary layer equations For a laminar boundary layer, the assumption of thinness leads to the Prandtl boundary-layer equations which, in the two-dimensional case, are

$$\frac{\partial u}{\partial t} + u\,\frac{\partial u}{\partial x} + v\,\frac{\partial u}{\partial y} = -\frac{1}{\rho}\,\frac{\partial p}{\partial x} + \mu\,\frac{\partial^2 u}{\partial y^2}$$

$$\frac{\partial u}{\partial x} + \frac{\partial v}{\partial y} = 0$$

Inspection shows that these equations are homogeneous in XY (and Z), and, indeed, it is the preservation of this homogeneity which mathematically determines the dimensions of p and μ. It follows that, provided no dimensional transformations are introduced, analysis in terms of XYZ may profitably be used in situations involving laminar boundary layers.

We now examine two examples illustrative of the foregoing principles.

7.2.1 Flow near an oscillating flat plate

It is required to find the fluid velocity $u(y, t)$ at time t and distance y from an infinite flat plate oscillating in its own plane. We consider the fluid to be incompressible and the motion is taken as laminar, being confined to, say, the x direction.

Let the velocity of the plate be given by $U = U_0 \cos(\omega t)$. The condition of no slip at the boundary will then imply that the velocity of the fluid at $y = 0$ coincides with that of the plate, and in consequence we have $u(0, t) = U_0 \cos(\omega t)$. Now since derivatives of u with respect to x and to z are clearly zero, it follows, in the light of our previous remarks, that the equations of fluid motion will be characterised by complete homogeneity and the analysis may be carried out in terms of $MXYZT$. We have then as the indicial matrix

	M	X	Y	Z	T
$u(y, t)$	0	1	0	0	−1
y	0	0	1	0	0
t	0	0	0	0	1
U_0	0	1	0	0	−1
ω	0	0	0	0	−1
ρ	1	−1	−1	−1	0
μ	1	−1	1	−1	−1

Solution of the indicial equations yields 3 DPs and we have

$$u/U_0 = \phi\, (y(\omega\rho/\mu)^{1/2},\, \omega t)$$

Dimensional analysis in terms of MLT would have resulted in 4 DPs and, say,

$$u/U_0 = \phi\, (\rho U_0 y/\mu,\, \omega y/U_0,\, \omega t)$$

and a consequent failure to maximise the amount of information available.

Finally, as a matter of record, the complete solution deriving from conventional analysis is given by

$$u/U_0 = \exp\left[-y(\omega\rho/2\mu)^{1/2} \cos(\omega t - y(\omega\rho/2\mu)^{1/2})\right]$$

7.2.2 Flow near a rotating flat disc

We next discuss incompressible, laminar flow around a flat disc of radius r rotating about an axis perpendicular to its plane with uniform angular velocity ω in an otherwise stationary fluid. We consider two associated problems involving, firstly, the determination of the velocity distribution within the fluid and, secondly, the moment required to maintain the motion of the disc.

Now physical independence clearly exists between the plane of the disc and the orthogonal direction but, in view of relationships such as $u = \omega r$, there will be no independence between orthogonal directions, say radial and tangential, in the plane of the disc itself. We choose, then, to work with the set $MLZT$, where L represents any direction in the disc plane and Z represents directions parallel to the disc axis. Working with the kinematic viscosity ν and directing our attention to u_r, the radial component of the fluid velocity, we have as the indicial matrix

	M	L	Z	T
u_r	0	1	0	-1
r	0	1	0	0
z	0	0	1	0
ω	0	0	0	-1
ν	0	0	2	-1
ρ	1	-2	-1	0

This yields 2 DPs and

$$u_r = \omega r \phi_1 (z(\omega/\nu)^{1/2})$$

For u_t, the tangential velocity component, the analysis is identical to that already carried out and we have

$$u_t = \omega r \phi_2 (z(\omega/\nu)^{1/2})$$

while for the axial velocity component we have

$$u_a = (\nu\omega)^{1/2} \phi_3 (z(\omega/\nu)^{1/2})$$

Analysis in the unextended MLT set would, of course, have given 3 DPs in each case. Reference may again be made to Schlichting[64], who treats the problem analytically and gives explicit expressions for each of ϕ_1, ϕ_2 and ϕ_3, the three functions that have been left undetermined above.

Coming now to the second half of our problem, we consider the moment M required to maintain the disc in motion.

The indicial matrix in this case is

	M	L	Z	T
M	1	2	0	−2
ρ	1	−2	−1	0
ν	0	0	2	−1
ω	0	0	0	−1
r	0	1	0	0

The matrix is, as expected, non-singular and gives one DP only. We find that

$$\frac{M}{\rho\omega^{3/2}r^4\nu^{1/2}} = \text{constant}$$

which is a complete solution.

Substitution of the relationship $U = \omega r$ in the expressions for the conventional dimensionless numbers C_M and Re shows that the solution obtained above may be written as

$$C_M = k \cdot Re^{-1/2}$$

where k is found by other methods to be equal to 3.87.

A further comment: had we carried out the analysis in terms of MLT, we should have arrived merely at the relationship $C_M = \phi(Re)$. This latter solution is applicable equally to laminar, transitional or turbulent flow, whereas our complete solution, based as it was on an extended set of length dimensions, is, for reasons which have been made clear, valid only in the case of laminar flow.

7.3   The dimensional representation of quantities and the role of 'hindsight'

It is advisable at this point to make a few additional observations concerning:

1. the choice of the set of reference dimensions appropriate to a particular problem, and

2. the assignment of a dimensional representation to the variables concerned.

As was seen in 5.3, the co-ordinate system in which we work may be cartesian, cylindrical, spherical, intrinsic or whatever may fit most naturally into the situation to be dealt with. The only basic criterion, apart from convenience, is that the system chosen must be such that dimensional homogeneity is maintained, and we now make the important point that transformation from one co-ordinate system to another, although frequently possible, will sometimes result in the breakdown of homogeneity, even though this may have previously been established. We shall, for example, see in 7.7 a case in which an intrinsic co-ordinate system is the only one in which homogeneity is preserved and which permits the problem to be treated by dimensional methods.

The transformation of a co-ordinate system and the concomitant transformation of the component length dimensions may also result in the development of a singularity in the indicial matrix or even in the impossibility of determining the dimensional representation of all variables present. It seems, then, that more detailed thought must be given to the question of the allocation of dimensions to certain variables when the dimensional set used has been extended beyond the basic MLT. We are, for instance, constantly faced with the difficulty of determining the dimensions of a 'characteristic' length which may feature in a particular situation. Here a modicum of physical insight or experience is necessary and this will often prompt us to choose, say, a diameter, rather than an area or circumference, as a variable characterising the given situation.

If the examination of a mathematically derived equation shows that a characteristic length has the dimensions $X^a Y^b$, we may, provided that the approach is otherwise valid, choose to work with two independent characteristic lengths of dimensions X and Y. It is granted that this will not result in any decrease in the number of DPs for, while the number of dimensions is increased by one, so also is the number of variables. The point we make, however, is that, if the conventional mathematical solution is known, then this technique may be used to gain greater physical understanding into precisely how variously orientated lengths enter into the situation and contribute to its restraints. Thus if a 'compound' characteristic length of dimension $X^a Y^b$ is brought to light, this should confirm or supplement our physical intuition concerning the conditions of the problem. The technique may then be used, as in **5.6**, to show the relative significance of the diameter and the circumference in pipe-flow problems.

But there are dangers involved in a direct dimensional analysis that has not been preceded by an analytical solution. If we are to place reliance upon our allocation of dimensions to a doubtful quantity, we must be very sure that the extended set of dimensions used is a valid basis for working in the situation which is being treated. An excellent example of the type of difficulty we have in mind, and in which the unwary are liable to be misled, arises from the problem of steady creeping flow past a sphere. Examination of the analytical solution shows that there is no dimensional homogeneity in XYZ. This results basically from the fact that the effect of viscosity is distributed over all three orthogonal directions and μ cannot, therefore, be given any unique dimensional representation. However, had physical independence erroneously been assumed and had the most obvious representation been ascribed to viscosity, that is $\mu \equiv MX^{-1} T^{-1}$ for flow in the x direction, then it is possible that we might have attempted a solution by working in XYZ. Since inertial effects in creeping flow are negligible and since a simple power solution for the drag force does, in fact, exist, we should have obtained the correct result, given by $C_D.Re = k$ reducing to $D = k.U\mu d$.

Now this result is consistent with the allocation of a characteristic length for the sphere having the dimension of X, and we are in consequence tempted to say that 'obviously' it is the diameter parallel to the direction of flow which is important and we might support our statement by pointing out that it is over this length that the quantity viscosity is effectively operative.

But, as already mentioned, a more detailed investigation indicates that there is in fact no homogeneity in XYZ and our deduction concerning the characteristic dimension of the sphere must, therefore, have been at fault. It also follows that we were in error in suggesting that the effect of viscosity is confined to the x direction. This well illustrates the need for caution against pitfalls deriving from too facile a use of extended dimensional systems in situations where their validity does not apply. Critical reading of Huntley [34] can be instructive in this respect.

The example just considered raises the question of 'hindsight', which may be defined as a tendency to look back after a solution has been obtained, saying, in effect, that, since our solution should be dimensionally homogeneous, this particular quantity must be dimensionally represented in this particular manner; and that, moreover, it is 'obvious' that this will be so. This tendency is clearly to be deprecated and leads to mistaken notions of what can be accomplished by dimensional analysis.

Nevertheless, it seems at least possible that, if we choose to regard dimensional analysis as an *ad hoc* tool which need not always be rigorously justified, then we may occasionally note that in restricted types of situation a given quantity invariably appears with a certain dimensional representation. The allocation of this representation to the quantity concerned may, then, lead to correct results in further situations of the same general type and which are worked in the same extended reference set. That this may be so is put forward as an idea only: it is not our intention to pursue it systematically.

As mentioned in **5.2** and discussed more fully in **10.6**, the existence of physical independence between two orthogonal lengths allows us to change the scale of one relative to the other or effectively to distort the situation without invalidating any results that have been obtained. While this is certainly the general case, there is, however, a danger in fluid dynamics that, although distortion appears permissible, it may nevertheless be of such a degree as to invalidate the conditions necessary for physical independence. This comment is particularly applicable to situations involving laminar boundary layers. An extended set of dimensions may, for example, validly be used to investigate the properties of the laminar boundary layer on a flat plate aligned with the stream direction, but if this plate is subsequently distorted into an elliptical or circular cylinder, then, for the reasons outlined in **7.2**, physical independence between orthogonal length dimensions no longer holds and the use of XYZ becomes invalid.

We also note that the nature of the distortion and its effects depend largely on the co-ordinate system used. Thus in a cartesian system, the linear distortion of an ellipse results in a second ellipse of different aspect ratio and the new body shape may result in the breakdown of dimensional homogeneity as mentioned above. In an intrinsic co-ordinate system, however, distortions will leave the boundary profile unaltered even though the geometry within the fluid itself will be transformed. This may, for example, result in a thicker boundary layer and, therefore, flow at a lower Reynolds number.

7.4   Laminar boundary layers and distortion

As mentioned in 7.2, important applications of the method involving the use of an extended set of length dimensions are to be found in laminar-boundary-layer theory. We proceed to consider in some detail the boundary layer on a flat plate aligned with the uniform free stream and we subsequently discuss the effect of distorting this into an elliptical or circular cylinder. A number of important results may be obtained with remarkable economy of effort.

7.4.1   Velocity distribution (flat plate)

We require to determine the velocity $u(x, y)$ at a point within the boundary layer, sited at a distance x downstream from the leading edge of the plate and y from its surface.

Figure 12

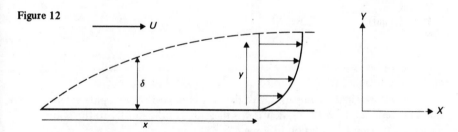

The free stream velocity is U, as shown in figure 12, and the indicial matrix will be

	M	X	Y	Z	T
$u(x, y)$	0	1	0	0	−1
x	0	1	0	0	0
y	0	0	1	0	0
U	0	1	0	0	−1
ρ	1	−1	−1	−1	0
μ	1	−1	1	−1	−1

This yields 2 DPs and we have

$$\frac{u(x, y)}{U} = \phi\left(y \cdot \left(\frac{U\rho}{\mu x}\right)^{1/2}\right)$$

This solution, which is constantly referred to in the literature has, thus, been derived with a pleasing minimum of calculation.

7.4.2   Boundary-layer thickness

Referring again to figure 12, we represent the thickness of the boundary layer at a point x by δ and we readily obtain a complete solution

$$\frac{\delta U^{1/2} \rho^{1/2}}{x^{1/2} \mu^{1/2}} = \text{constant}$$

or, more simply,

$$\delta/x = k \cdot Re^{-1/2}$$

7.4.3   Local and overall friction

τ_x, the local shear stress at x, and the overall shear stress, $\dfrac{1}{x}\displaystyle\int_0^x \tau \cdot dx$, will each be dependent upon ρ, μ, U and x. Analysis gives one DP and we have for each case

$$\frac{\tau^2 x}{\rho U^3 \mu} = \text{constant}$$

and, accordingly,

$$C_f \text{ (local)} \quad = k_1 \cdot Re^{-1/2}$$
$$C_F \text{ (overall)} \quad = k_2 \cdot Re^{-1/2}$$

Note that C_f and C_F are equidimensional, the latter being regarded as the mean value of the former up to the point x.

This solution may be compared with that for laminar flow in a pipe where $x (\equiv X)$ is replaced by $d (\equiv Y)$, that is d is measured orthogonally to the boundary. In such a situation the solution becomes:

$$\frac{\tau d}{\mu U} = \text{constant}$$

or $C_f = k \cdot Re^{-1}$

The position is shown in figure 16, which also relates to turbulent flows and will be discussed further in 7.5.

7.4.4   Drag on plate

Here again the drag force per unit width of plate D' will be dependent upon ρ, μ, U

and x, giving one DP, and in consequence we have

$$C_D = k \,.\, Re^{-1/2}$$

7.4.5 Skin-friction drag on elliptical cylinder

With these results behind us, we are now free to progress from flat plates to the treatment of an ellipse with the major axis aligned in the free stream direction. It will be of particular interest to regard this as an example involving distortion and we may take as our two extreme cases the flat plate, already considered, and the circular cylinder. The most readily to accommodate this distortion we decide to work with an intrinsic co-ordinate system, as indicated in figure 13.

Figure 13

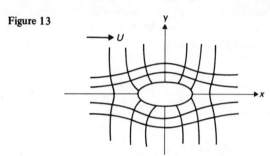

Over the upstream portion of the ellipse, where the boundary layer is thin, the methods already developed may be applied to find an expression for the local shear stress set up at the elliptical surface. However, at some distance back from the stagnation point, corresponding to the forward end of the major axis, the boundary layer widens and separation quickly occurs. It follows that analysis in terms of XYZ may no longer be used. Nevertheless, the region of the widening boundary layer is of limited extent and does not contribute significantly to the overall situation. Moreover, the shear stresses in the wake, that is in the region downstream from the separation points, are wholly negligible and it follows that a suitable approximation for the skin friction drag (i.e. the drag due to shear stress alone) may be obtained by considerations confined to the thin boundary layer only.

Using the notation indicated in figure 14, we argue for any particular

Figure 14

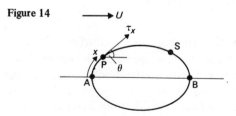

(elliptical) cylinder as follows. At any point P on the surface before the onset of separation at S and at a distance x from the forward stagnation point A, the local shear stress τ_x will be dependent on ρ, U, x and μ and we have, therefore

$$\tau = k \cdot x^{-1/2} \rho^{1/2} U^{3/2} \mu^{1/2} \tag{1}$$

this result being obtained in the same manner as in **7.4.3**.

But we are primarily concerned with the total drag due to skin friction considered for the cylinder as a whole, and this will be given by

$$D' = 2 \int_A^B \tau_x \cdot \cos \theta \cdot dx$$

where D' represents the drag per unit width of the cylinder. The integration is carried out between the forward stagnation point and its antipodal counterpart, but, but, since shear stresses in the wake are negligible, integration need, in fact, only be taken as far as the separation point and we have then

$$D' = 2 \int_A^S \tau_x \cdot \cos \theta \cdot dx$$

We now use equation 1 in order to rewrite this as

$$D' = k \cdot \rho^{1/2} U^{3/2} \mu^{1/2} \int_A^S x^{-1/2} \cos \theta \cdot dx \tag{2}$$

We proceed by arguing that the location of the separation point S will be given by $x_s = f(\rho, U, d, \mu)$, where d is a characteristic length of the cylinder measured in the x direction. It follows that x_s will be independent of Re and is, indeed, fixed for any particular ellipse. We have, that is to say: x_s/d = constant. Since, then, we have shown that the location of the separation point is dependent only on the geometrical properties of the ellipse, the integral contained in equation 2 may consequently be replaced by $k \cdot l^{1/2}$, where k is an undetermined numerical constant and l is a characteristic length, say the major diameter. We are now able to write equation 2 in the form

$$D' = k \cdot \rho^{1/2} U^{3/2} \mu^{1/2} l^{1/2}$$

and thus

$$C_D = k \cdot Re^{-1/2}$$

We have then been successful in showing that the drag due to skin friction on a cylinder is inversely proportional to the square root of the Reynolds number − a result which holds up to fairly high values of Re and is subject only to the boundary layer remaining laminar. We have, however, said nothing of the pressure drag or the total drag. Indeed, the total drag is proportional to Re^{-1}, but this holds only for low values of Re where inertial effects are negligible.

7.5   Turbulent flows

In this section an acquaintance with elementary aspects of turbulent flow is assumed. Reference may, in case of need, be made to Bradshaw[4] or to Schlichting[64].

As a preliminary we mention that the dimensional analysis of turbulent flows makes widespread use of the ratio $(\tau_w/\rho)^{1/2}$ where τ_w is the wall or boundary shear stress. This quantity has the dimensions LT^{-1} and is known as the 'friction velocity' u^*. We shall ourselves be making use of this quantity, but we remind the reader that any argument in which u^* occurs may readily be rephrased in such a way as to depend directly upon τ_w.

It will be recalled that, in considering laminar flow (7.4) we took the boundary shear stress as a dependent variable and carried out the analysis in terms of U, which was used as an independent variable characterising the flow. In contrast with this approach, we tend when working with turbulent flows, to regard τ_w, and therefore u^*, as having greater physical significance if it be used as an independent variable, and this approach will be reflected in the examples which follow. Indeed, we note that the use of dp/dx in equation 1 of 5.6 was wholly equivalent to the procedure now proposed. There is, however, no necessity about our recommendation and the method used is largely governed by considerations of convenience and convention.

7.5.1   Uniform turbulent flow in smooth circular pipe

In order to illustrate these points, we consider in some detail the commonly occurring situation involving uniform turbulent flow through a smooth pipe of circular cross-section.

We have already treated the problem of finding the wall stress under given flow conditions and we recall from (7.4) that τ_w is dependent upon ρ, U, d and μ. This dependence may be reduced by suitable analysis to

$$\frac{\tau_w}{\frac{1}{2}\rho U^2} = \phi\left(\frac{\rho d U}{\mu}\right)$$

or $C_f = \phi(Re)$

a relationship which is equally well expressed in terms of u^*.

If now we seek to determine the velocity distribution within the pipe, we may proceed with the analysis by noting that $u(y)$ will be dependent upon ρ, u^*, a, y and μ, where y denotes distance from the pipe wall and a is the pipe radius. This gives the relationship:

$$u/u^* = \phi\,(Re^*_y, y/a) \tag{1}$$

where $Re^*_y = u^* y/\nu$, that is a Reynolds number based on the friction velocity and with y included explicitly as the length component. $\nu = \mu/\rho$ is, of course, the kinematic viscosity.

Near the pipe wall, where a is not expected to enter directly into the problem, we have simply:

$$u/u^* = \phi\,(Re^*_y) \tag{2}$$

a relationship commonly referred to as the 'law of the wall'.

At the pipe axis, where u maximises to U_m and $y = a$, we have from equation 1

$$U_m/u^* = \phi(Re^*_a) \tag{3}$$

And finally in the region near the pipe axis, where turbulent stresses are dominant and where the effects of viscosity make only an insignificant contribution to variations in velocity, we have:

$$(U_m - u)/u^* = \phi\,(y/a) \tag{4}$$

This is known as the 'velocity defect law', since the quantity $(U_m - u)$ expresses the velocity defect in the region.

The argument may be taken a stage further by considering an overlapping region where both equations 2 and 4 apply. Thus we have from equation 2

$$\frac{u}{u^*} = \phi_1\left(\frac{u^* y}{\nu}\right) = \phi_1\left(\frac{u^* a}{\nu} \cdot \frac{y}{a}\right)$$

while from equation 4 we have

$$u/u^* = U_m/u^* - \phi_3\,(y/a)$$

$$= \phi_2\,(u^* a/\nu) - \phi_3(y/a) \quad \text{(by virtue of equation 3)}$$

Equating the right-hand sides now gives

$$\phi_1\,(u^* a/\nu \cdot y/a) = \phi_2\,(u^* a/\nu) - \phi_3\,(y/a)$$

The solution of this type of functional equation must clearly be logarithmic and it follows that in the overlapping region the 'law of the wall' may be written as

$$u/u^* = A \cdot \ln\,(u^* y/\nu) + B \tag{5}$$

This last equation will not, however, hold in the immediate vicinity of the wall since the turbulent fluctuations, being damped out, result in viscous stresses dominating the situation and, in consequence, the 'law of the wall' here reduces to

$$u/u^* = u^* y/\nu \tag{6}$$

While this is equivalent to the assumption of a linear velocity distribution, it may perhaps be better derived by the direct use of the extended set of dimensions, since the condition of orthogonal independence holds in the laminar sub-layer under discussion (7.2).

The results given by equations 5 and 6 are illustrated in figure 15.

Figure 15 Law of the wall: $u/u^* = \phi'(u^*g/v)$

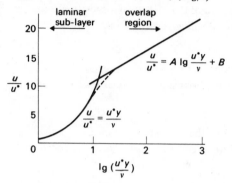

If it be further assumed that the 'log law' applies to the mean velocity U, we may rewrite equation 5 as

$$U/u^* = A \cdot \ln(u^*d/v) + B \tag{7}$$

By resorting now to the defining equations

$$Re = U \cdot 2a/v \qquad C_f = \tau_w/\tfrac{1}{2}U^2\rho \qquad u^* = (\tau_w/\rho)^{1/2}$$

we may express equation 7 in the form

$$C_f^{-1/2} = A \cdot \ln(Re \cdot C_f^{1/2}) + B$$

where the constants A and B are not necessarily the same as those occurring in equation 5. This form of equation is found to agree well with experimental results which are shown set out in figure 16.

7.5.2 Turbulent boundary layer and roughness

The turbulent layer on a flat plate may be considered in much the same manner. In this new situation we have merely to replace the pipe radius by the boundary-layer thickness and the maximum pipe velocity by the velocity outside the boundary layer. With these substitutions we derive precisely the same general laws as for the case of turbulent flow through a circular pipe.

If the preceding analysis be extended to take into consideration the effects of roughness, it is clear that a further variable will enter into the situation. This will be the roughness parameter k, a length that characterises the magnitude of the roughness projections, and its inclusion will result in the introduction of one further DP which will usually be represented by (u^*k/v) or by (k/y).

Figure 15 shows that for, say $(u*k/v) < 5$, the projections are within the viscous sublayer and the roughness is expected to have little effect upon the flow. In consequence it is said that under such conditions the surface is 'hydraulically smooth'. For $(u*k/v) > 100$, however, the roughness projections extend into the region of turbulence and result in major modifications to the flow pattern. The surface is then termed 'hydraulically rough' and the effects of viscosity are found to be insignificant. The velocity distribution will accordingly take the form

$$u/u* = A \cdot \ln{(y/k)} + B$$

The effects of roughness on the friction coefficient may be appreciated by an examination of figure 16. For this situation, dimensional analysis readily yields

$$C_f = \phi(Re, k/d)$$

Note, in particular, that the figure brings out the manner in which C_f becomes dependent upon k/d and independent of Re in certain ranges. Duncan, Thom and Young[16] may be consulted for further details. ●

Figure 16 Effects of roughness coefficient: $C_f \phi (Re, k/d)$

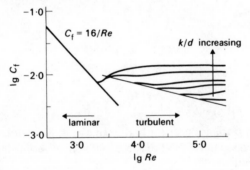

7.6 Compressible flows

Where conditions are such that the compressibility of the fluid becomes a significant factor, then a variable which takes this into account must be introduced into the dimensional analysis. This will generally be the bulk modulus K and the associated new DP will be $(U/(K/\rho)^{1/2})$, which we recognise as the Mach number. The denominator here has the dimensions of velocity and will be equal to the speed of sound a, since $a = (K/\rho)^{1/2} = (\partial p/\partial \rho)^{1/2}$. The Mach number, then, is usually defined in terms of the velocity ratio (U/a) and is based on the characteristic velocity U which is usually taken as the free stream velocity.

A number of additional considerations have to be made when compressibility becomes significant. Thus the quantities K and ρ may be liable to change throughout the flow, and the concept of a local Mach number is accordingly used. A further independent DP will then be necessary, this generally being the specific-heat ratio γ. In many compressible flow problems, viscosity may be largely ignored, but where it does warrant consideration, the fluid motion will generate heat and the Prandtl number ($\mu c_p/\kappa$) must be taken into account.

Some types of flow may lead to large temperature variations, free convection or other special processes, and their consideration will then involve, for example, use of the Nusselt, Grashof or other dimensionless numbers. No detailed discussion will be given here but the interested reader may refer to the many specialist texts available. (See, for example, Liepmann and Roshko[42].)

A further and more general point. Considerable use is made of the Laplace and similar equations in the investigation of fluid flows. Now it was pointed out (7.2) that, where a Laplace type equation holds, it is not possible to regard XYZ as physically independent. Nevertheless, it is common to introduce distortion successfully into fluid dynamical problems and, for instance, to determine the flow past an elliptical cylinder from a knowledge of the flow past a circular one. In order to avoid confusion we are led to make the important observation that, while a physical independence is certainly a sufficient condition for the legitimate use of distortions involving a linear scaling factor applied to one of the orthogonal directions of the co-ordinate system, it is by no means a necessary condition for the more general types of 'functional' distortion made use of in theoretical fluid dynamics.

● 7.7   **Appendix on similar flows**

We consider the problem of similar flows — that is to say flows in which the velocity profile is identical, except possibly for a linear scaling factor, over any cross-section of the boundary layer.

It will be recalled from **7.3** that, in considering the velocity distribution in the neighbourhood of a flat plate, we found u to be dependent upon U, ρ, μ, x and y. Dimensional analysis then gave

$$\frac{u}{U} = \phi \left(y \left(\frac{U}{\nu x} \right)^{1/2} \right)$$

Now, in order that any other flow is to be considered 'similar', we shall require a result of the form $u/U = \phi(\text{DP})$ where 'DP' represents one dimensionless product only, since a relationship of this nature will define a particular velocity profile, whereas the emergence of any additional DP will, in general, permit the profile to vary with x.

The determination of such flows has been studied from the point of view of conventional mathematical analysis by Falkner and Skan[23] and by Goldstein[30], but we here outline a particularly simple approach based on the ideas of orthogonal independence which have been developed.

We find it convenient to use an intrinsic system and to define a characteristic velocity U as that of the free stream beyond the influence of boundaries. Bearing in mind the fact that our ideas on the nature of the flow involved are not yet fixed, we assume simply that u, the velocity component in the x direction of the flow at the point (x, y) will be dependent upon x, the distance from a defined point on the boundary and measured along the boundary; y, the distance measured orthogonally to the boundary; U, the characteristic velocity already defined; ρ and μ.

Regarding u as the dependent variable we find, as before, that

$$\frac{u}{U} = \phi\left(y \cdot \left(\frac{U}{\nu x}\right)^{1/2}\right)$$

If, however, we include among our list of independent variables a characteristic length d associated with the boundary profile, we shall then have an additional DP and

$$\frac{u}{U} = \phi\left(yy \cdot \left(\frac{U}{\nu x}\right)^{1/2}, \frac{d}{x^a y^b}\right)$$

where $a + b = 1$, this relationship holding whatever the dimensions of d in terms of X and Y may be.

It will be clear from the introductory remarks that in order to obtain a solution involving similar flow d can have no place in the argument of ϕ and there can, in consequence, be no characteristic length with which the boundary may be associated. By a straightforward consideration of the geometry of the situation we conclude that the only boundary fulfilling this condition will be either a straight line or the intersection of two (or more) straight lines at a point.

A simple demonstration of this is as follows: a boundary profile may, in general, be defined by $y = \Sigma a_n x^n$, where n may take any values positive and/or negative. Now if this equation is to be dimensionally homogeneous in L, we see that a_n will be dimensionless only for the case of $n = 1$. It follows that it is possible to define (at least) one characteristic length unless a_n is held zero for all values of n other than $n = 1$. We then have as our boundary profile simply $y = ax$, a straight line or, by an extension of our argument, any number of straight lines intersecting at a point.

Visualisation of the position is simple. Consider the boundary formed by two straight lines intersecting at an angle α. For $\alpha = 0$ we have the flat-plate solution already discussed. For $0 < \alpha < \pi$ we have the so-called wedge flows as indicated in figure 17, where the hatched area represents the 'solid behind the boundary' inaccessible to the fluid. Where $\alpha = \pi$, we have two-dimensional stagnation flow.

Figure 17

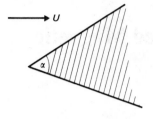

Where $\pi < \alpha < 2\pi$ we have a channel, convergent or divergent according to the direction of U.

Where $\alpha = 2\pi$, we have no space for the fluid, or, alternatively a pipe or channel with parallel sides. Uniform flow taking place within such a channel will, of course, represent a special case of similar flow. It is true that there will here be a characteristic length d, the breadth of the channel, but as there will be no well-defined characteristic point, x inevitably drops out of the equation. Dimensional analysis leaves us with two groups only as given by

$$u/U = \phi\,(y/d)$$

thus confirming that the flow remains similar.

For many values of α the flows mentioned may be difficult to achieve in practice. Thus flow into or out of a 'wedge' channel ($\alpha > \pi$) will clearly involve a point sink or source. This represents a singularity, and flow in its immediate neighbourhood would not be covered by the argument. Again, in practice, if the angle of the channel divergence be too great, the flow may undergo separation.

It has nowhere been explicitly suggested in the foregoing discussion that the direction of U has to be parallel to the axis of the wedge or channel, and no theoretical restrictions are laid down in the argument concerning the orientation of U. ●

8

Dimensions of Electric and Magnetic Quantities

8.1 Introduction

As in the case of thermal quantities, a variety of approaches is possible when dealing with electric and magnetic quantities. Each approach depends upon the sequence and the manner in which the fundamental definitions are made and, provided that these are developed consistently, no one approach can be considered as 'right' at the expense of the others being 'wrong'. Historically, the systems most commonly used have been:

1. the electromagnetic system in which the reference dimensions are $MLT\mu$, where μ is the dimension of permeability;

2. the electrostatic system, in which the reference dimensions are $MLT\epsilon$, where ϵ is the dimension of permittivity.

With the introduction and general adoption of the SI there is a case for bringing the conventionally accepted dimensional set into conformity with the SI units. This would lead to 'current' as the fourth 'electrical' dimension, corresponding to the ampere, which is the basic electrical unit (**1.3**). Current, however, is so clearly a derived quantity ($i = \mathrm{d}q/\mathrm{d}t$), that we prefer, arbitrarily perhaps but in accordance with a growing practice, to standardise upon 'charge', with dimension Q as our reference quantity and to work, therefore, in the $MLTQ$ system. This approach has a particular advantage in that its use results in the dimensional representation of the majority of electrical quantities being characterised by integral exponents. The position is much the same where current is used as a dimension but, as a glance at Table A3 (**page 205**) will show, fractional exponents proliferate in both the $MLT\mu$ and the $MLT\epsilon$ systems, suggesting perhaps that these represent, in some ways, a less 'natural' basis for analysis.

It is clear that there are no logical, as opposed to aesthetic, objections to a preponderance of fractional indices occurring in dimensional formulae and although, for example, $M^{1/2}$ may have no physical meaning, it is nevertheless a respectable mathematical entity upon which we may operate without risk of ambiguity. Be this as it may, the fact that the adoption of *MLTQ* leads to neatness and clarity of analysis prejudices us in favour of its use. This will not, however, inhibit us from resorting to other systems on an *ad hoc* basis when situations arise to which they may be particularly suited.

8.2   The *MLTQ* system

The deduction of the dimensions of electrical quantities is a straightforward enough exercise. We take electric charge q as the one new undefined reference quantity and, using this as a base, we select a series of defining equations in such a fashion as will lead most elegantly to the introduction of the various derived electrical quantities and to the development of the theory as a whole.

We have, for instance, very simply,

Electric current is defined as $i = dq/dt$ and has the dimensions $i \equiv T^{-1}Q$.

Potential difference between two points is defined as the work done in moving a unit charge from one point to the other and has, in consequence, the dimensions $V \equiv ML^2 T^{-2} Q^{-1}$.

Resistance is defined as $R = V/i$ and has, therefore, the dimensions $R \equiv ML^2 T^{-1} Q^{-2}$.

Permeability The dimensions of μ may most conveniently be derived from Ampère's equation giving the force per unit length between long parallel wires separated by a distance d and carrying currents i_1 and i_2. This is $f/l = \mu i_1 i_2/2\pi d$ and it follows that we have the representation $\mu \equiv MLQ^{-2}$.

Permittivity. Using the equation for the force between two charged particles, $f = q_1 q_2/\epsilon d^2$, we find that $\epsilon \equiv M^{-1} L^{-3} T^2 Q^2$.

Note that in deriving the dimensions of μ we used Ampère's equation, being unable to resort to $f = m_1 m_2/\mu d^2$ in view of the fact that the dimensions of magnetic pole strength m have still to be determined in the particular sequence of definitions adopted. It is clear that the one equation cannot be used to establish the dimensions of two new quantities. However, now that the dimensions of μ in *MLTQ* have been derived, we are entirely free to use $f = m_1 m_2/\mu d^2$ in order to ascertain the dimensions of magnetic pole strength; to maintain homogeneity here we have

Magnetic pole strength: $m \equiv ML^2 T^{-1} Q^{-1}$

We mention in passing that such equations as

$$V = L\frac{d^2 q}{dt^2} + R\frac{dq}{dt} + \frac{q}{C}$$

applying to a circuit containing an inductance L, a resistance R and a capacitance C in series; or

$$Z = R^2 + \left[\left(\omega L - \frac{1}{\omega C} \right)^2 \right]^{1/2}$$

giving the impedance Z of a circuit containing a resistance, inductance and a capacitance in series, are useful for checking the dimensional consistency of the various quantities involved. This applies equally, of course, to the $MLTQ$ and to all alternative systems.

There are no particular difficulties in the procedure outlined and for ease of reference the dimensional structure of a selection of electrical and magnetic quantities will be found listed in Table A3 (page 205).

8.3   The electromagnetic and electrostatic systems: $MLT\mu$ and $MLT\epsilon$

In our discussion we use μ and ϵ indiscriminately to denote the permeability and permittivity of both empty space and of a medium. This procedure helps simplify our argument and is justified in view of the fact that the dimensions of each of the quantities concerned are not affected by where that quantity is measured. Our interest in the electromagnetic and the electrostatic systems is largely due to the historical importance which they have played in the development of dimensional theory, and we proceed, therefore, to a brief consideration of certain points that invite attention.

The basic equation of the electromagnetic system is $f = m_1 m_2 / \mu d^2$, giving the force between two magnetic poles. This at once provides the dimensions of magnetic pole strength in the $MLT\mu$ system as $m \equiv M^{1/2} L^{3/2} T^{-1} \mu^{1/2}$. A theoretical objection sometimes raised against this approach is that no isolated magnetic pole has yet been found to exist in nature; it is suggested, therefore, that it is incorrect to use this concept in a defining equation. Nevertheless, there are a number of reasons which suggest that isolated poles may exist and, indeed, research is currently being carried out in the attempt to locate this elusive monopole (Ford[26]). We will not attempt to prejudge the issue, and it will be sufficient to point out that the magnetic pole is a satisfactory concept which is given precise meaning by virtue of the relationship between electrical and magnetic effects. If it eventually turns out to have no physical analogue, this should not discredit its use in determining the structure of units; and, since the approach involving the use of μ as one of the reference set of dimensions leads to no inconsistencies, this may be regarded as sufficient reason for its adoption.

Returning to our main theme, we have also to consider the electrostatic system and we see that this may be founded directly upon the basic equation $f = q_1 q_2 / \epsilon d^2$ which gives the dimensions of charge in the $MLT\epsilon$ system as $q \equiv M^{1/2} L^{3/2} T^{-1} \epsilon^{1/2}$.

Further derivations will not be carried out in either system, as they represent a routine exercise.

There does, however arise a point which has generated interest and discussion in the past. In the basic equation of the electromagnetic system, $f = m_1 m_2 / \mu d^2$, we were faced with two quantities m and μ neither of which appears directly expressible in terms of MLT only. It seemed, then, a natural approach to select one of these quantities as being a new and physically independent reference dimension and to express the other in terms of it. Our choice for the reference quantity happened to be μ. Similarly in the electrostatic system, the two new quantities which appear to be physically independent of MLT are q and ϵ. In this system, then, we opt for the latter quantity, taking ϵ as the new reference dimension.

We ask, however, whether μ and ϵ will invariably be physically independent of MLT or whether it may not be possible in certain situations to allocate to them the dimensions of the MLT system. In much the same way we have already seen that temperature θ, although generally independent of MLT, may occasionally be expressed as $\theta \equiv ML^2 T^{-2}$.

In considering this we note first that we have no more (and no less) right to assume that μ is dimensionless in $f = m_1 m_2 / \mu d^2$ or ϵ dimensionless in $f = q_1 q_2 / \epsilon d^2$ than we have to assume that G is dimensionless in $f = G . m_1 m_2 / d^2$ (3.3). Indeed it was explicitly stated by Rücker[60] as far back as 1889 that μ and ϵ cannot *both* be dimensionless. If this were the case it would be implied that:

$$f \equiv \frac{m_1 m_2}{d^2} \equiv \frac{q_1 q_2}{d^2}$$

and that in consequence the dimensions of m and of q would each be $M^{1/2} L^{2/3} T^{-1}$. This is unacceptable as it leads to obvious inconsistencies, as may be seen from a consideration of Ampère's equation quoted in 8.2 and which, with μ dimensionless, would then yield $q \equiv M^{1/2} L^{1/2}$.

It is, nevertheless, readily possible to express the product of μ and ϵ in terms of MLT. This result may be deduced from the dimensional representation of these quantities as listed in 8.2. Multiplying together the two representations gives the result $1/\mu\epsilon \equiv L^2 T^{-2}$.

One of the many alternative methods of deriving this is by a direct examination of the electromagnetic wave equation:

$$\frac{\partial^2 E}{\partial t^2} = -\frac{1}{\mu\epsilon} \cdot \frac{\partial^2 E}{\partial x^2}$$

Here the quantity E is the electrical field intensity, but its dimensions do not for the present concern us in that they cancel out leaving simply $1/\mu\epsilon \equiv L^2 T^{-2}$ as before.

A further approach of some interest is due to Carr [9], who pointed out that

$f = q_1 q_2 / e d^2$ strictly applies only to stationary charges. For non-stationary charges we invoke the special theory of relativity and transform the equation into

$$f = \frac{1}{ec^2} \cdot \frac{(q_1 v_1)(q_2 v_2)}{d^2} \cdot \sin \alpha$$

where v_1 and v_2 are the charge velocities, α is the angle between the directions of their movement and c is the velocity of light. This reduces to

$$f = \frac{1}{ec^2} \cdot \frac{ii' \cdot ds\ ds'}{d^2}$$

and, comparing this with Ampère's equation, we see that $1/\mu\epsilon = c^2$ and $1/\mu\epsilon$ is therefore again shown to have the dimensions of (velocity)^2.

We have still to obtain a representation of μ and of ϵ separately in MLT, and there appears to be no elementary situation in which the necessary physical dependence occurs. Perhaps the most direct approach is to equate the dimensions of the Bohr magneton, $he/4\pi mc$, with those of magnetic moment. In the $MLT\mu$ system the quantities comprising the magneton are

Planck's constant	$h \equiv ML^2 T^{-1}$
electronic charge	$e \equiv M^{1/2} L^{1/2} \mu^{-1/2}$
electronic mass	$m \equiv M$
velocity of light	$c \equiv LT^{-1}$

Equating the dimensions of the PP formed from these to those of

magnetic moment $M \equiv M^{1/2} L^{5/2} T^{-1} \mu^{1/2}$

gives $\mu \equiv L^{-1} T$

Since we have already shown that $1/\mu\epsilon \equiv L^2 T^{-2}$, it follows that in the MLT system the dimensions of both μ and ϵ will be those of velocity and $\mu \equiv \epsilon \equiv L^{-1} T$.

There is no inconsistency with what has been said concerning μ and ϵ as reference dimensions. Our interpretation relates specifically to the situation being considered and, provided this situation be such that μ and ϵ are physically independent of MLT, a gain in sharpness of analysis will result from acknowledging the fact and by extending the reference set accordingly.

We also note that we may make use of the equivalence $\mu \equiv \epsilon \equiv L^{-1} T$ by making a suitable substitution in the dimensional representation of either μ or ϵ as already derived in the $MLTQ$ system. We find then that $Q \equiv ML^2 T^{-1}$, and this will be the dimensional form of Q in situations in which a physical dependence between electrical and mechanical quantities may arise.

We will ourselves make little use of the $MLT\mu$ and $MLT\epsilon$ systems, although either one of them may conveniently be adopted in situations which are characterised, respectively, by a predominance of either magnetic or electrical quantities (see 8.5.1 and 8.5.2).

A number of other systems have also been suggested. Thus in electrical engineering the concept of mass is not used extensively and it is, therefore, often convenient to work with electric potential Φ as a reference dimension in place of M. Furthermore, since electrostatic fields may be of small importance in heavy-engineering applications, the dimension ϵ may be replaced by the dimension of current I. This leads to the $LTI\Phi$ system in which mass has the dimensions $M \equiv L^{-2} T^3 I$. This approach may be adapted readily enough to a variety of techno-logical problems, but it will not be considered further since its use leads to no particular difficulties of interpretation.

8.4 Rationalised units

One point should be mentioned before proceeding to examples. This involves the question of rationalised units. It will have been noted that we have expressed the equations for permittivity and permeability as

$$f = \frac{q_1 q_2}{\epsilon d^2} = \frac{m_1 m_2}{\mu d^2}$$

As is well known, this 'natural' approach results in the factor 4π frequently occurring in the equations of electromagnetic theory even though it may appear anomalous by reason of the fact that there is no spherical symmetry.
Similarly the factor is liable to be absent where such symmetry exists. Oliver Heaviside was impressed by this when, in 1899, he wrote:

> The unnatural suppression of 4π in the formulas of central force, where it has a right to be, drives it into the blood, there to multiply itself, and afterwards break out all over the body of electromagnetic theory.

In order to obviate this aesthetic defect, various 'rationalised' systems of units have been adopted over the years and are, indeed, now incorporated in the SI. The rationa-lised form of an equation, then, may readily be obtained by, for example, substituting $\mu/4\pi$ and μ and $4\pi\epsilon$ for ϵ, and this will usually have the effect of considerable simplification. Thus the unrationalised equation for the capacity of a parallel plate condensor is

$$C = k \cdot \frac{\epsilon A}{d} = \frac{1}{4\pi} \cdot \frac{\epsilon A}{d}$$

whereas, in the rationalised form, this reduces to

$$C = \frac{\epsilon A}{d}$$

The consideration of units is outside the general scope of this book, but we raise it here since it may be objected that, as a result of the procedure of rationalisation, the constants of a physical equation undergo a change and that this is in contradiction to the position previously established for dimensionally homogeneous equations (2.3). The apparent paradox may, however, readily be resolved. If in the above equation we make the simple substitution of $4\pi\epsilon$ for ϵ and leave the matter at that, then clearly the value of k will be affected; indeed, a change in k is the whole object of the rationalisation. If, however, we work consistently in the system chosen, say here the $MLT\epsilon$ system, and if we accordingly make corresponding changes in the magnitude of the units in which C is measured and as indicated in the dimensional representation $C \equiv L\epsilon$, then k will of course remain unchanged. We have, then, no exception to the rule that the value of a DP will be unchanged for any conversion of the units — provided that the conversion be systematically applied to each of the quantities appearing.

More generally it follows that when we rationalise we change the units of μ and ϵ and we should, if we wish to be consistent, make corresponding changes in the units of all quantities defined in terms of μ and ϵ. Provided this be done, both rationalised and unrationalised forms of the equation will be homogeneous and any constants appearing in them will be invariant — but this would defeat the whole object of rationalisation.

The procedure actually adopted, then, makes it inevitable that the values of the units of certain quantities will vary between rationalised and unrationalised systems. This can be a source of confusion and, if it be insisted that the units of all quantities are to remain unchanged irrespective of the system used, it will be entailed that the equations of one or other of the systems will have to be regarded as numerical rather than physical. The current tendency is to choose units in such a way that the rationalised equations are regarded as homogeneous and the unrationalised forms have, in consequence, to be regarded as numerical. This, however, need not concern us, as from the dimensional point of view the difficulty is an artificial one. The reader who wishes to pursue the matter further may, however, consult Ipsen[35], where the question is considered in more detail.

8.5   Examples

We consider a few examples of dimensional analysis applied to the field of electricity and magnetism. Where appropriate, the use of XYZ in this book will from now on be made without it being considered necessary to raise the question of rigorous justification.

8.5.1   Radius of curvature of an electron in a magnetic field

We take the situation involving an electron projected at right angles to a uniform magnetic field. Since this problem clearly involves magnetic field strength, we feel that the $MLT\mu$ system may be the most suitable. In order to minimise the number of variables, we decide initially to work with ω, the angular velocity of the electron, as the dependent variable. Independent variables will then be the mass of the electron m, the magnetic field strength H, the charge on the electron e and the permeability of the medium μ. These quantities give as our indicial matrix

	M	L	T	μ
ω	0	0	−1	0
m	1	0	0	0
H	$\frac{1}{2}$	$-\frac{1}{2}$	−1	$-\frac{1}{2}$
e	$\frac{1}{2}$	$\frac{1}{2}$	0	$-\frac{1}{2}$
μ	0	0	0	1

Analysis now gives one DP and

$$\omega = k \, . \, \mu e H/m$$

To rewrite this in terms of radius of curvature we make use of the result $\omega = v/\rho$ and, in consequence, we obtain the required solution:

$$\rho = k \, . \, \frac{mv}{eH\mu}$$

8.5.2 Force between charged condenser plates

Since this problem involves permittivity, we decide to work in the $MLT\epsilon$ system. The dependent variable is the force f, while the other variables are the distance between the plates d, their area A and the voltage difference between them V. This gives the following matrix:

	M	L	T	ϵ
f	1	1	−2	0
d	0	1	0	0
A	0	2	0	0
ϵ	0	0	0	1
V	$\frac{1}{2}$	$\frac{1}{2}$	−1	$-\frac{1}{2}$

In view of the linear dependence that has arisen between the M and the T columns, the rank is only 3 and there are in consequence 2 DPs which yield the equation

$$f = V^2 \, \epsilon \phi \, (A/d^2)$$

A complete solution is only to be arrived at by making the possibly obvious assumption that the force is proportional to the plate area, in which case we have

$$f = k \cdot V^2 \, \epsilon A/d^2 \tag{1}$$

An alternative approach to the same problem may be made if we take advantage of the fact that there is physical independence between the X dimension taken as normal to the plates and the L dimension representing directions lying in the planes of the plates. We choose this time to work with the dimension Q, instead of ϵ, and accordingly we have in the $MXLTQ$ system

	M	X	L	T	Q
f	1	1	0	−2	0
d	0	1	0	0	0
A	0	0	2	0	0
ϵ	−1	−1	−2	2	2
V	1	2	0	−2	−1

Commenting upon this, we see that the dimensions of ϵ, being a scalar property of the medium between the plates, will be symmetrical in the three orthogonal length dimensions, whereas the voltage difference enters into the situation only by reason of its effect in the x direction and is consequently to be represented by the dimensions shown.

We now have only one DP and, without having to fall back upon any assumptions concerning the proportionality of force and area, we write immediately

$$f\left(\frac{fd^2}{V^2 \epsilon A}\right) = 0$$

which reduces to equation 1 above.

Working more rigorously, we may avoid an appeal to the principle of symmetry in deriving the dimensions of ϵ by the following argument. In the defining equation of ϵ, that is $f = q_1 q_2/ed^2$, we note that d^2 is, in effect a measure of the area normal to the line joining the point charges and it follows that the dimensional representation of d^2 should be written as L^2. Indeed the position is precisely similar to that which we experienced in 5.6 where we found that the numerical equivalence between A and r^2 was not, in fact, a dimensional equivalence and that something was therefore lost by the substitution $A = 2\pi r^2$ which yielded a dimensionally heterogeneous equation in the extended system. Bearing this in mind, we return to the defining equation for ϵ and find that the dimensions of permittivity do indeed conform to the representation given in the matrix.

8.5.3 The skin effect

High-frequency currents are largely confined to a thin layer at the surface of the conductor. This is the so-called 'skin effect'. The thickness d of the conducting skin is defined to be such that its specific resistivity to a steady current would be equal to the observed specific resistivity at the frequency considered. We have, then, to determine d as a function of the frequency n, the permeability of the medium μ, the radius r and the resistivity ρ of the conducting wire. We use the $MLTQ$ system which gives the matrix

	M	L	T	Q
d	0	1	0	0
n	0	0	−1	0
μ	1	1	0	−2
r	0	1	0	0
ρ	1	3	−1	−2

There is an unexpected singularity here, since the Q column is −2 times the M column. We have, therefore 2 DPs and

$$f(d/r, \rho/nr^2\mu) = 0$$

or $d = r\,\phi\,(\rho/nr^2\mu)$

If we now assume that the skin depth is sufficiently thin for it to be independent of the wire radius, we see that ϕ must be equal to $(\rho/nr^2\mu)^{1/2}$ and we consequently obtain a complete solution:

$$d = k \cdot \left(\frac{\rho}{n\mu}\right)^{1/2}$$

We may be struck by the formal similarity of this solution to that of **6.3.3** and, indeed, both the phenomenon now under consideration and that of the penetration of temperature fluctuations into the subsoil are governed by the same type of partial differential equation, the factor (ρ/μ) in the present electromagnetic problem being equivalent to the factor $(\kappa/\rho c)$ which occurred in the thermal problem. Again, the situation involving laminar flow in the neighbourhood of an oscillating flat plate (**7.2.1**) may be taken as yet another physical analogue.

8.5.4   Oscillation of a LCR circuit

We require the period of oscillation of a circuit containing an inductance L, a capacitance C, and a resistance R. Analysis yields

$$t = k \cdot \frac{L}{R}\,\phi\left(\frac{CR^2}{L}\right)$$

For a simple series or parallel circuit it may be shown that R has a negligible effect upon the period and we have then $\phi = (CR^2/L)^{1/2}$ with, in consequence,

$$t = k \cdot (LC)^{1/2}$$

An identical analysis may be applied in order to determine the time-constant t of a transient circuit containing the same elements. Specialisations of the result may readily be obtained and we see that

$t = k_1 \cdot RC$ for negligible L

$t = k_2 \cdot L/R$ for negligible C

$t = k_3 \cdot (LC)^{1/2}$ for negligible R

8.5.5 Energy of an electromagnetic field

We wish to determine the space energy u in electrostatic and electromagnetic fields. Simple analysis shows at once that, in an electrostatic field only, u will be proportional to the permittivity ($u = k \cdot \epsilon E^2$), while in a magnetic field only it is proportional to the permeability ($u = k \cdot \mu H^2$).

We proceed to enquire what we may learn concerning the value of u in the general electromagnetic field. Choosing (arbitrarily) the $MLT\mu$ system to work in, we prepare the matrix

	M	L	T	μ
u	1	-1	-2	0
E	$\frac{1}{2}$	$\frac{1}{2}$	-2	$\frac{1}{2}$
H	$\frac{1}{2}$	$-\frac{1}{2}$	-1	$-\frac{1}{2}$
ϵ	0	-2	2	-1
μ	0	0	0	1

This is singular, since $M \equiv L^{-1/3} T^{-1/3}$, and of rank 3, giving 2 DPs from which we find:

$$u = \epsilon E^2 \phi \left(\frac{\mu H^2}{\epsilon E^2} \right)$$

In order that this result may conform with the two special results already obtained and relating to the cases in which the magnetic field and the electrostatic field respectively are zero, we are tempted to write

$$\phi = k_1 + k_2 \frac{\mu H^2}{\epsilon E^2}$$

giving

$$u = k_1 \epsilon E^2 + k_2 \mu H^2$$

which is, in fact, the correct result with $k_1 = k_2 = 1/8\pi$, but dimensional analysis alone gives us no justification for saying so.

This example provides an excellent illustration of the way in which a singularity develops in the indicial matrix as a direct result of the fact that the solution to the problem considered is not expressible in the form of a PP (11.3).

8.5.6   Ohm's law

We conclude with a more sophisticated example. In the attempt to deduce Ohm's law from dimensional considerations we prepare the table:

Physical quantity	Symbol	M	L	T	Q
Current density	J	0	−2	−1	1
Electric field strength	E	1	1	−2	−1
Number of free electrons per unit volume	N	0	−3	0	0
Charge on electron	e	0	0	0	1
Mass of electron	m	1	0	0	0
Electronic mean free path	λ	0	1	0	0
Mean velocity of electrons	v	0	1	−1	0

We decide to work with current density and electric field strength rather than with current and voltage, as this allows us to reduce the number of variables present. Our approach involves an appreciation that the wire length l enters the problem only via the relationship $V/l = E$, and the cross-sectional area of the wire only via $i/A = J$. Here we have the 3 DPs contained in the equation

$$f\left(\frac{J}{Nev}, \frac{eE}{mv^2\lambda}, N\lambda^3 \right) = 0$$

We next introduce an auxiliary argument in order to render the solution complete. We note firstly that it is the drift velocity of the electrons which produces the current and it follows that we may reduce the number of variables further by working with this drift velocity v_d rather than with both J and N, that is we make use of the equation $J = v_d eN$. Secondly, we see that the number of reference dimensions may be increased by one if we distinguish between the X dimension corresponding to the direction of current or electron drift and the L dimension corresponding to all directions orthogonal to this. The distinction enables us to work in the extended $MXLTQ$ system and we have as our matrix

	M	X	L	T	Q
v_d	0	1	0	−1	0
E	1	1	0	−2	−1
e	0	0	0	0	1
m	1	0	0	0	0
λ	0	$\frac{1}{3}$	$\frac{2}{3}$	0	0
v	0	$\frac{1}{3}$	$\frac{2}{3}$	0	0

A few comments will be appropriate:

1. The drift velocity v_d has the dimensions XT^{-1} since it is drift in the x

direction that results in the current.

2. The dimensions of electric field strength must, then, similarly be expressed in terms of X.

3. On the other hand, the mean velocity of electrons and the mean free path must have their length dimensions expressed symmetrically since these quantities will be evenly distributed in all three orthogonal directions.

We are now in a position to obtain a complete solution

$$f\left(\frac{v_d mv}{eE\lambda}\right) = 0$$

or, upon substitution of our relationship for v_d and rearranging,

$$E = k.\left(\frac{mv}{Ne^2\lambda}\right).J$$

which is equivalent to Ohm's law.

The reader interested in the subtleties of our subject should contrast this approach with that used in **6.4.1** where we considered the thermal conductivity of a gas. In this previous problem there was no overall drift of molecules in the direction of heat transfer and we were concerned, rather, with the rate of penetration through the gas and in the direction of the temperature gradient of the increase in the mean random velocity of the molecules. It was, therefore, the components of the mean free path and of the mean molecular velocity in the x direction that were significant, and their dimensional representations were reflected accordingly. In the present problem involving current flow, the random electron velocity, as opposed to the drift velocity, is unchanged as a result of the passage of the current.

There is an interesting extension to the foregoing argument which introduces a consideration of both thermal and electrical conductivities in metals. Note first that electrical conductivity, defined as J/E, may be expressed as

$$\sigma = k_1 \frac{Ne^2\lambda}{mv}$$

Now if we consider the free electrons to form, in effect, a monatomic gas, we may make use of the result for thermal conductivity obtained in **6.4.1**. This was

$$\kappa = k_2 . mNv\lambda c_V$$

The ratio of thermal to electrical conductivity will then be given by

$$\frac{\kappa}{\sigma} = k.\frac{mc_V}{e^2}. mv^2 \tag{1}$$

Recalling from the kinetic theory of gases the relationships

$$mc_V = k_1 \Re$$

and $$mv^2 = k_2 \, \Re \, \theta$$

we substitute these in equation 1 to obtain

$$\frac{\kappa}{\sigma} = k \cdot \left(\frac{\Re}{e}\right)^2 \theta$$

This final equation expresses the Lorenz law which states that 'the ratio of thermal to electrical conductivity is proportional to the absolute temperature' and also the Wiedmann—Franz law which states that at a particular temperature this ratio maintains the same value for different metals (Joos[36]).

9

The Reduction of Undetermined Functions

9.1 The determination of additional relationships between variables

A basic problem in dimensional analysis lies in determining how to conduct the argument in such a way that the number of DPs is reduced to a minimum and preferably to one. Only thus will it be possible to reduce the degree of functional indeterminacy occurring in the solution and, in consequence, to maximise the precision of the solution that becomes available.

We have seen in Chapter 5 that functional indeterminacy may often be reduced by increasing the number of dimensions which comprise the reference set. As a further approach to this same problem we now concern ourselves with the possibility of placing certain types of restriction upon the selected list of variables. We may, for example, be able to introduce an *a priori* relationship between the variables and in so doing limit the manner in which the various DPs enter into the solution, thus effectively reducing their number. In certain cases, moreover, the new relationship may even be useful in eliminating one or more of the variables and consequently we may directly reduce the number of DPs that are thrown up.

As a special case involving this approach, we shall consider the possibility of minimising the number of variables by noting that we are able to treat some combination of them as a single quantity. An instance of this occurs when we regard as a significant variable the weight of a body, $W = mg$, rather than the two separate quantities m and g, or when we decide to work with kinematic viscosity v rather tha. with the two variables μ and ρ ($v = \mu/\rho$).

The introduction of a new relationship into an analysis is liable to involve a degree of insight into a situation over and above the basic appreciation of what variables are relevant. It involves an approach that has already been adopted in a large number of examples in the text. (For example, in 5.3.1, where we regarded an angle of deformation as proportional to a couple, and in 8.5.6, where we made a number of assumptions concerning relationships involved when a current passes through a wire.) We now attempt to demonstrate the possibilities in a more systematic fashion.

In general we consider a situation in which the relationship between the significant variables is

$$f(a, b, c, \ldots, p, q, r, \ldots) = 0 \tag{1}$$

which reduces to

$$f(\pi_1, \pi_2, \ldots, \pi_n) = 0 \tag{2}$$

There will then be three obvious methods by which the number of variables may be effectively decreased and by which a consequent reduction may be made in the number of DPs occurring in equation 2.

Method 1 Suppose first that we have prior knowledge of an additional dimensionally homogeneous equation connecting a sub-set of the variables occurring in equation 1. Let us write this as

$$p = \phi(q, r, \ldots) \tag{3}$$

Clearly, then, whenever the quantity p occurs in equation 1 it may be simply replaced by $\phi(q, r, \ldots)$, with a consequent reduction in the number of effective quantities and a corresponding reduction in the number of DPs which appear in equation 2.

Note that this approach is not limited to the introduction of a single additional relationship but, within obvious limitations, may be extended to embrace the case of two or more such equations.

For illustrative purposes we take an example involving the introduction of two new relationships, each of the form of equation 3, and we proceed to investigate the period t of a conical pendulum which we take to be dependent upon the length of the string l, the height of the cone h, the gravitational acceleration g, the angular velocity ω and the inclination of the string to the vertical α. We have, then, corresponding to equation 1,

$$f(t, l, h, g, \omega, \alpha) = 0 \tag{4}$$

which involves 6 variables and in the LT system yields 4 DPs. In this case, however, we are able to introduce two additional equations each corresponding to equation 3:

$$\left. \begin{array}{l} h = l \cos \alpha \\ \omega = 2\pi/t \end{array} \right\} \tag{5}$$

Since in the final solution h may invariably be replaced by $l \cos \alpha$ and ω by $2\pi/t$, we may clearly reduce equation 4 to

$$f(t, l, g, \alpha) = 0$$

Use of the two additional relationship has, then, resulted in a decrease in the original number of variables by 2 and we are left, in consequence, with 2 DPs only, namely $(t^2 g/l)$ and (α), these leading to the solution

$$t = (l/g)^{1/2}$$

The introduction of additional homogeneous relationships to effect a reduction in the number of DPs may also be accomplished after the main dimensional analysis has been completed. Suppose, starting from equation 1 we have reached equation 2 and subsequently decide to make use of equation 3. We note that equation 3 may itself be written in the form $\pi^*_1 = \phi(\pi^*_2, \ldots)$, where the π^*'s are a set of DPs based upon $p, q, r, \ldots$ only. But equation 2 may readily be brought into a form which includes this set of π^*'s and it follows that, since $\pi^*_1 = \phi(\pi^*_2, \ldots)$, π^*_1 may be omitted from the (amended) argument of equation 2. The required reduction in the number of DPs has then been carried out subsequently to the completion of the initial dimensional analysis.

Method 2 Our second method involves an extension of Method 1 and becomes applicable when the situation is such that we have reason for believing that the variables $p, q, r, \ldots$ appear *only* in the form of equation 3. If such be the case, instead of merely deleting p from the indicial matrix, we may delete $q, r, \ldots$ and retain in their place only the (compound) variable p.

Reverting to our example of the conical pendulum, we recall that we introduced relationships in equation 5. If we are able to supplement this introduction with the statement that l, α and ω do not appear in equation 4 except as shown in equations 5, then each of l, α and ω may be deleted from our list of variables and we obtain a complete solution:

$$t = k(h/g)^{1/2} \quad \text{or} \quad t = k \cdot (l \cos \alpha/g)^{1/2}$$

In general it is seldom possible to take advantage of such statements, since the grounds for making them are not usually present.

Method 3 The additional relationship to be introduced need not necessarily be homogeneous and, indeed, in the most common cases we resort to what is merely an empirical requirement of proportionality which we suppose to exist between two of the variables.

We frequently have reason to expect that a variable p will be directly or indirectly proportional to another variable q $(p = k \cdot q^{\pm 1})$ and we arrange our DPs in such a fashion that p occurs only in π_1 and q occurs only in π_2. If the requirement of proportionality is to be satisfied, it will then be immediately obvious how π_1 and π_2 must be combined to form a single DP.

Numerous examples of this approach have been seen in the foregoing text. We mention here the condenser problem (8.5.2) where the solution involved two DPs and, in effect, we had

$$f(\pi_1, \pi_2) = f\left(\frac{f}{V^2\epsilon}, \frac{A}{d^2}\right) = 0$$

Using the additional nonhomogeneous proportionality relationship $f = k \cdot A$, it follows that π_1 and π_2 must be combined in the form

$$\pi_1 \cdot \pi_2^{-1} = \frac{f}{v^2\epsilon} \times \left(\frac{A}{d^2}\right)^{-1} = \text{constant}$$

and we immediately have the complete solution:

$$f = k \cdot \frac{V^2\epsilon A}{d^2}$$

We proceed to consider one or two further problems the successful solution of which depends upon the determination of simple *a priori* relationships between the variables.

9.1.1   Deflection of a cantilever

We investigate the deflection d of a cantilever beam of length l and of rigidity R under the influence of an applied weight W. The indicial matrix is

	M	L	T
d	0	1	0
W	1	1	-2
l	0	1	0
R	1	3	-2

Since this is of rank 2 only, it yields 2 DPs and, as a partial solution, the somewhat uninformative

$$d = l \, \phi \, (R/Wl^2)$$

A very limited insight, however, suggests that small deflections will be proportional to the weight and we are, in consequence, able to introduce the simple, nonhomogeneous relationship $d = k.W$ as an additional restriction upon the solution. From this it follows that ϕ can only be of the form $\phi = (R/Wl^2)^{-1}$ and the final and complete solution may be written

$$d = k \cdot Wl^3/R$$

9.1.2 *Terminal velocity of a liquid drop*

A less obvious problem illustrating the same principles lies in the investigation of the terminal velocity of a liquid drop falling through air. We prepare the following table:

Physical quantity	Symbol	M	L	T
Terminal velocity of drop	U	0	1	−1
Drop radius	r	0	1	0
Gravitational acceleration	g	0	1	−2
Air viscosity	μ	1	−1	−1
Air density	ρ_1	1	−3	0
Liquid density	ρ_2	1	−3	0

The usual approach leads to the functional equation:

$$f\left(\frac{U^2}{gr}, \frac{\rho_1 Ud}{\mu}, \frac{\rho_2}{\rho_1}\right) = 0 \tag{1}$$

where the second DP will be recognised as the Reynolds number.

As it stands, this result is not informative, but it may be rendered fully determinate by the following argument in which we adopt a direct approach, the formal justification for which lies in our discussion of Method 1. For small velocities, where inertial effects are negligible, the resistance to the falling drop will, on dimensional grounds, be given by $f = k_1.\mu Ur$. Again, on dimensional grounds, the weight of the drop is given by $W = k_2.\rho_2 r^3 g$ and the upthrust due to buoyancy by $B = k_2 .\rho_1 r^3 g$. Now it is evident that when the drop attains its terminal velocity it will have no acceleration and its weight will exactly counteract the sum of the resistive and buoyancy forces. Equating these we have

$$k_1 . \mu Ur = k_2(\rho_2 - \rho_1)r^3 g \tag{2}$$

and hence the complete solution

$$U = k .\frac{r^2 g}{\mu}(\rho_2 - \rho_1) \tag{3}$$

Commenting upon this we note that we effectively introduce two new (homogeneous) relationships between the variables. Firstly the difference in densities replaces the two individual densities and secondly, since the resistive force is equated to the apparent weight, we obtain equation 2 and it follows, for instance, that $U \propto g$, when all other quantities remain constant. By reason of these relationships, the three DPs in equation 1 reduce firstly to

$$f\left(\frac{U^2}{gr}, \frac{(\rho_2 - \rho_1) Ud}{\mu}\right) = 0$$

and subsequently to

$$f\left[\left(\frac{U^2}{gr}\right).\left(\frac{(\rho_2 - \rho_1)\,Ud}{\mu}\right)^{-1}\right] = 0$$

and we see that the original function given in equation 1 is equivalent to

$$f\left[\left(\frac{U^2}{gr}\right).\left(\frac{\rho_1\,Ud}{\mu}\right)^1.\left(\frac{\rho_2}{\rho_1}-1\right)^{-1}\right] = 0$$

which reduces immediately to equation 3.

9.1.3   The fan laws

A further, if more involved, example of the principles we are discussing lies in the deduction of the series of results known to engineers as the 'fan laws'. Our approach here is based on Hamilton, Kusian and Cope[32].

As a preliminary we obtain an equation applicable to turbulent flow of air through pipes and known as Atkinson's law. Provided that the effects of air viscosity and compressibility are negligible, the pressure drop Δp in a pipe may be expressed in terms of the air density ρ, the volume flow rate Q, the cross-sectional area A and the length l, giving as the indicial matrix

	M	L	T
Δp	1	−1	−2
ρ	1	−3	0
Q	0	3	−1
A	0	2	0
l	0	1	0

Analysis yields

$$\Delta p = \frac{\rho Q^2}{A^2}\,\phi\left(\frac{l^2}{A}\right)$$

We now take the fall in pressure along the pipe to be proportional to the distance travelled by the air, that is to l, and we have $\phi\,(l^2/A) = k\,.\,(l^2/A)^{1/2}$, from which we immediately deduce Atkinson's law in the form

$$\Delta p = k\,.\,\frac{\rho Q^2 l}{A^{5/2}} \tag{1}$$

With this lemma behind us, we may consider the question of axial flow fans. The variables affecting the (static) pressure developed across a fan Δp will be Q, the fan diameter d, the air density ρ and the angular velocity of the fan n. The matrix, in consequence, will be

	M	L	T
Δp	1	-1	-2
Q	0	3	-1
d	0	1	0
ρ	1	-3	0
n	0	0	-1

Analysis here yields

$$\frac{\Delta p}{\rho n^2 d^2} = \phi_1 \left(\frac{Q}{nd^3}\right) \tag{2}$$

We next regard the power input of the fan P as dependent upon the same set of variables and, in this case,

$$\frac{P}{\rho n^3 d^5} = \phi_2 \left(\frac{Q}{nd^3}\right) \tag{3}$$

We proceed to obtain a further important independent variable, the efficiency η, from equations 2 and 3 by writing

$$\frac{\Delta p}{\rho n^2 d^2} \cdot \frac{\rho n^3 d^5}{P} \cdot \frac{Q}{nd^3} = \frac{Q}{nd^3} \cdot \phi_1 \left(\frac{Q}{nd^3}\right) \cdot \phi_2 \left(\frac{Q}{nd^3}\right)$$

Now $\Delta p.Q$ is the 'gas static power output' and the (static) efficiency of the fan may then be written as

$$\eta = \frac{\Delta p \cdot Q}{P} = \phi_3 \left(\frac{Q}{nd^3}\right) \tag{4}$$

Let us apply these results to the case where the size of both fan and duct is fixed, that is the case of a 'constant system'. We then have

1. the pressure loss along the duct is equal to the static pressure Δp developed by the fan (assuming that the duct is open-ended), and

2. A and l remain constant.

It follows from equation 1 that $\Delta p = k \cdot \rho Q^2$, a result which we substitute in equation 2 to give

$$k \cdot \frac{Q^2}{n^2 d^2} = k \cdot d^4 \left(\frac{Q}{nd^3}\right)^2 = \phi_1 \left(\frac{Q}{nd^3}\right)$$

leading to

$$k \cdot d^4 = \phi_4\left(\frac{Q}{nd^3}\right)$$

Since d is held fixed in a 'constant system', it follows that (Q/nd^3) is constant and we have, from equations 2, 3 and 4:

$$\left.\begin{aligned}
\frac{Q}{nd^3} &= \text{constant} \\[2mm]
\frac{\Delta p}{n^2 d^2} &= \text{constant} \\[2mm]
\frac{P}{n^3 d^5} &= \text{constant} \\[2mm]
\eta &= \text{constant}
\end{aligned}\right\} \tag{5}$$

From these results, in which the proportionality constants may not be dimensionless, we may deduce such fan laws as *for constant air density and for constant system*:

$$\left.\begin{aligned}
Q &\propto n \\
\Delta p &\propto n^2 \\
P &\propto n^3 \\
\eta &= \text{constant}
\end{aligned}\right\}$$

Alternatively, if we like to consider various geometrically similar fans operating at a 'fixed point of rating', that is with η held constant, then from equation 4 we see that (Q/nd^3) will continue as a constant and equations 5 will, therefore, still be applicable. We deduce, then, such further laws as *for geometrically similar fans operating at a fixed point of rating, constant air density and constant tip speed (nd = constant), we have*

$$\left.\begin{aligned}
Q &\propto d^2 \\
\Delta p &= \text{constant} \\
P &\propto d^2
\end{aligned}\right\}$$

The whole basis of our success has lain in the introduction of restrictive relationships between subsets of the variables and it is, therefore, an excellent example of the techniques we have been discussing. A point of special interest is that we started by considering the variables Δp, P, Q, n, d and ρ. Had we treated the problem as a whole we should have obtained $(n - r) = (6 - 3) = 3$ DPs and, in consequence,

$$f\left(\frac{\Delta p}{\rho n^2 d^2} \ , \ \frac{P}{\rho n^3 d^5} \ , \ \frac{Q}{nd^3}\right) \ = 0$$

from which no useful result could be deduced. We proceeded, therefore, by basing our analysis upon the restrictive assumption that Δp, and P were each separate dependent functions of Q, n, d and ρ. This resulted in equations based on 2 DPs only, thus paving the way for our subsequent introduction of Atkinson's law, which represented a further restriction upon the possible relationships existing between the variables.

9.2   The combination of variables and the avoidance of ratios

A particular, and one of the more commonly occurring, aspects of the approach considered in 9.1 lies in the possibility of considering two or more variables entering the analysis as a single group. Thus in 6.3.2 we worked with the quantity $(g\beta)$, treating it as a single variable and thus reducing by one the number of DPs produced. Quite commonly we work with a mass flow rate $\dot{m}$ rather than with the two variables m and t, and, again, in problems of beam flexure we make frequent use of the compound quantity (EI), since we know that in certain types of situation the two quantities E and I invariably occur as a product.

We make two comments. Firstly, there is no need for the combination of variables to be multiplicative. (Recall that in 9.1.2 we worked with the single variable $(\rho_2 - \rho_1)$ rather than with ρ_1 and ρ_2 separately.) Secondly, when working with a combination of quantities we have to satisfy ourselves that none of the individual quantities is liable to appear in its own right and divorced from the combination which we have opted to work with. This last requirement may not always be too easy to satisfy and may be subsumed formally under Method 2 of 9.1.

We also make a brief reference to the question of ratios. A half century ago there was a tendency to work with quantities such as 'specific gravity' defined as the ratio of the mass of the material considered to that of an equal volume of water, or 'specific heat' defined as the ratio of the thermal capacity of a substance to that of an equal mass of water. Quantities of this nature are clearly dimensionless and, instead of combining with one of the DPs that result from an analysis, simply enter into the solution as an additional DP with a consequent loss of information.

While mentioning this point we do not wish to labour it, for the contemporary student of engineering and physics is seldom tempted to adopt this approach and rightly prefers to work with quantities associated with a definite dimensional structure. Certain quantities, however, such as strain or Poisson's ratio, are 'naturally' expressed as ratios in that no reference is made to the characteristics of any arbitrary material in their definition. Where such quantities arise, it will generally be necessary to accept the position rather than to adopt some artificial subterfuge in the attempt to avoid it.

9.3 The decision to regard the effect of a variable as negligible

The decision to regard the influence of one or other of the variables in a problem as negligible will often throw light upon the nature of an undetermined function and may represent a more subtle approach than the mere excision of that variable from the list of quantities considered.

9.3.1 Drag on body immersed in a viscous incompressible fluid

The following example treats the classical problem of the drag on a body immersed in a viscous, incompressible fluid. Here we have

Physical quantity	Symbol	M	L	T
Drag (force)	D	1	1	−2
Velocity	U	0	1	−1
Linear dimension of body	d	0	1	0
Density of fluid	ρ	1	−3	0
Viscosity	μ	1	−1	−1

Analysis gives 2 DPs and we have

$$D = \rho U^2 d^2 \cdot \phi \, (U d \rho / \mu) \tag{1}$$

where the argument of the undetermined function is the Reynolds number. There are three cases to consider:

1. At very low velocities we may expect inertial forces to be negligible, with a result that drag will be almost wholly due to viscosity. ρ may therefore be regarded as negligible, implying that $\phi = (U d \rho / \mu)^{-1}$ and in consequence

    $$D = k \cdot \mu U d$$

2. At considerably higher velocities but before the onset of transition to a turbulent boundary layer, experimental evidence shows that the effect of ρ dominates the situation and μ has very little influence. (This applies generally within the range of $4 \cdot 0 < \lg Re < 5 \cdot 0$) Here, then, we may expect that $\phi = (U d \rho / \mu)^0$ and in consequence

    $$D = k \cdot \rho U^2 d^2$$

3. By making an assumption of continuity, we may expect that throughout the intermediate range the exponent of the Reynolds number will pass through all values between −1 and 0. Thus, for example, in the neighbourhood of some point we would have $\phi = (Ud\rho/\mu)^{-1/2}$ and the drag would then be given by

$$D = k \, . \, (\mu\rho U^3 d^3)^{1/2}$$

These three cases suggest that equation 1 may now be written as:

$$D = k \, . \, \rho U^2 d^2 (Ud\rho/\mu)^{\alpha}$$

where the value of α varies with the velocity range considered. We are, then, able to subsume in the one equation the effect of drag upon a submerged body for a wide range of velocities provided only that $\lg Re < 5$. (For high velocities where a turbulent boundary layer develops or where compressibility of the fluid becomes significant, the situation is of a fundamentally different nature and another approach to the analysis must be made.)

In confirmation of the results obtained let us refer to figure 18 which illustrates

Figure 18 Drag as a function of velocity

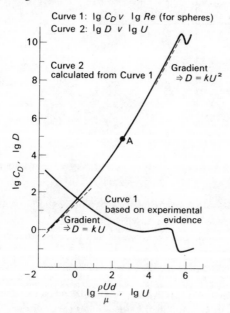

in a slightly idealised form the well-known empirically obtained plot of $\lg C_D$ against $\lg Re$. (It will be recalled that C_D is the drag coefficient, defined as $D/\frac{1}{2}\rho U^2 d^2$, and reference may be made to Goldstein[30].) As we are currently concerned with the dependence of D upon U we may regard ρ, μ and d as constant and curve 1 in the figure will then correspond to $\lg (D/U^2)$ plotted against $\lg U$. From the relationship $\lg D = \lg (D/U^2) + 2 \lg U$, a curve of $\lg D$ plotted against $\lg U$ may be simply derived and is also shown in the figure as curve 2.

Note that the gradient of this second curve for $\lg Re \leqslant 1$ is equal to unity, corresponding to $D \propto U$ as in case 1 discussed above. For $4\cdot0 \leqslant \lg Re \leqslant 5\cdot0$ the gradient is 2, corresponding to $D \propto U^2$ as in case 2. And finally the expected transition occurs between these two ranges with, for instance, $D \propto U^{3/2}$ in the neighbourhood of point A and corresponding to case 3.

As a variation on the technique of regarding certain quantities as negligible, it may be possible to regard them as held constant. An example of this approach leading to a reduction in indeterminacy was given in **6.3.1**, where we dealt with a situation in which it was considered justifiable to hold constant the Prandtl number.

One further point. We have, in this section, considered the possibilities of regarding quantities as negligible, despite the fact that they do enter into a physical situation. It will also be appropriate to enquire into the position if, through lack of insight, we take as significant a quantity that does not in principle find any place in the relationship which we seek to determine. We recall, for example, that in **2.5** we considered listing mass as a significant variable in the problem of the simple pendulum whereas, in fact, it was of no consequence. Again, in **6.4.1** we listed pressure as a variable despite the fact that it was not significant in the situation considered. Where inadvertently we consider an entraneous variable, one of two alternatives will generally occur. The first is that the superfluous variable drops out, as was the case in the two examples quoted; when this happens the solution to the indicial equations shows that it can only occur in a DP with zero power and no difficulty results. Alternatively, the superfluous variable may bring about the introduction of an additional DP which itself enters into the solution with zero power. This last possibility is liable to confuse the issue as we may have no means of knowing that such a situation has arisen and information will, in consequence, be lost.

9.4   Staicu's 'general dimensional analysis'

Staicu[65] has developed an approach to dimensional analysis of surprising power and simplicity, and this may often lead to a reduction in the degree of functional indeterminacy which would otherwise be present. The approach, however, is based upon the following three highly restrictive requirements which limit the scope of its application. It is assumed that:

1. the solution to the problem can be expressed as a PP;

2. the precise set of variables entering into the problem is known (the inclusion of an irrelevant variable leads, not to loss of information, but to positive error);

3. it is known whether the effect of an increase in the magnitude of each of the independent variables, taken individually, leads to an increase or a decrease in the magnitude of the dependent variable.

Staicu points out that, if the final result is to be a PP, it will be possible to express express the solution with integral exponents in the form

$$P^p \cdot L^l M^m N^n \ldots = k \cdot A^a B^b C^c \ldots \qquad (1)$$

where P is the dependent variable and $LMN \ldots$ and $ABC \ldots$ are the independent variables. $LMN \ldots$ are such that an increase in any one of them will effect a decrease in the magnitude of P, while $ABC \ldots$ are variables with the reverse effect. The exponents $p, lmn \ldots, abc \ldots$ are all positive integers.

The next step is to express the variables in terms of the r members of the reference set of dimensions and to write out the r (or fewer) indicial equations resulting from the fact that equation 1 will be homogeneous in each of these dimensions.

Staicu finally seeks the minimum positive integral values of the unknowns in these equations, starting with the equation with the least number of terms, provided that this exists. The value of the unknowns so determined provides the required solution.

Before commenting upon this, we illustrate the approach with one of Staicu's own examples. This seeks to determine the deflection s of a beam simply supported at the ends, having a length l and under the influence of a concentrated load P applied at a point located at distances x and y from the two ends. With E representing the elastic modulus and I the moment of inertia, equation 1 takes the form

$$s^a \cdot E^b I^c l^d = k \cdot P^e x^f y^g \qquad (2)$$

Working with the reference set MLT and writing the variables in dimensional form we find

$$(L)^a \cdot (L^{-1} M T^{-2})^b (L)^d \equiv (LMT^{-2})^e (L)^f (L)^g$$

The conditions for homogeneity in L and in M are then, respectively;

$$a - b + 4c + d = e + f + g \qquad (3)$$

$$b = e \qquad (4)$$

Note that here the equation for homogeneity in T will be identical with that for M, since T is not here an essential dimension, being physically dependent upon M with, in fact, $M \equiv T^{-2}$. Since equation 4 has fewer terms than equation 3, we solve it for minimum positive integers to find

$$b = e = 1$$

Equation 3 then gives:

$$(a + 4c + d) - (f + g) = 2$$

Here the condition for the allocation of minimum positive integral values entails that $a = c = d = 1$, giving

$$(a + 4c + d) = 6$$

and, in consequence,

$$f + g = 4$$

or $f = g = 2.$

The final results are then

$$a = b = c = d = e = 1$$

$$f = g = 2$$

and, with these values, equation 2 yields

$$s = k \,.\, Px^2 y^2 \,(EIl)^{-1}$$

which corresponds to the result obtained by conventional analytical methods.

Although this is Staicu's own example, it well illustrates the main shortcomings of his approach, and we make the following observations:

1. It may not have been obvious at the start of the analysis that s increases both with x and with y and that it decreases with l.

2. Had we contented ourselves with the length variables l and x only, and had we failed to include the 'redundant' y, the results would not have been expressible as a PP and one of the assumptions upon which Staicu's approach is founded would not have held good. Had we failed to appreciate this and continued nevertheless with the analysis, we should have obtained the erroneous result

$$s = k \,.\, Px^4 \,(EIl)^{-1}$$

instead of the correct

$$s = k \,.\, Px^2 (1 - x)^2 \,(EIl)^{-1}$$

3. The solution of the indicial equation $f + g = 4$ is indeterminate. Staicu selects $f = g = 2$, but there appears no good reason why he should not have selected $f = 3, g = 1$.

Despite the fact that Staicu's methods are dangerously liable to mislead, the reader is recommended to work a few examples for himself and, in general, he will come upon the correct result with an almost uncanny economy of effort.

Following Staicu's approach, there seems little advantage in extending the number of reference dimensions in which a problem is worked. Indeed, we may without loss of power generally work in a reduced set — even, in the limit, with only one dimension obtained, say, by replacing each of M, L and T by the nominal dimension X, provided that the problem be then formulated in such a way as to avoid the occurrence of variables such as velocity which would be dimensionless in X. This may, however, generally be accomplished by some such subterfuge as is involved in considering the two variables 'length' and 'time' instead of the one composite variable 'velocity'. Admittedly, however, this technique is liable to accentuate the difficulties of treating the indeterminate equations that arise. Thus Poiseuille's problem worked with one dimension and using the two variables m and t instead of the composite mass flow rate $\dot{m}$ gives the indicial equation $a + b + c = 6$, with nothing to indicate that the required solution of this is $a = b = 1$ and $c = 4$.

The reason for the pragmatic success of Staicu's minimum-integer law appears to be related to the empirical observation that physical quantities are simply defined and simply combined. In an attempt to illustrate this we selected at random, from a number of texts, a total of fifty equations expressible in the form of PPs. These were written in a form involving integral indices (for example, $t = k \cdot (l/g)^{1/2}$ was rewritten as $t^2 = k \cdot l/g$). The indices were then classified according to their magnitude and the following frequency count resulted:

Magnitude of index:	±1	±2	±3	±4	±5	±6
Frequency of occurrence:	157	33	10	4	0	0

This confirms that, provided a result is expressible as a PP, it will generally happen that any one variable contained within that result will occur with an index ±1 and it is progressively less likely that a second, third or higher power will be present.

The mathematical manipulation of the defining equations of physics simply does not lead to the proliferation of high indices. And Staicu, in his assumption that a result is expressible as one DP, by that very fact ensures conditions in which his 'minimum-integer law' is likely to lead to a correct solution. It follows that, while Staicu's approach does in general give a correct result, exceptions do occur, though admittedly they are not easy to find. For counter-examples the reader may be referred to equations 1 and 2 of **9.6.2**.

A further comment. The more information we put into a problem, the greater will be the precision which we should expect in the result. In straightforward dimensional analysis, it is easy to quantify this and if a result is based on one DP it may be taken as having unit precision. A result based on k DPs will then have the lesser precision of $1/k$ and, as we have seen, we may reduce k by the introduction of increased information resulting from, say, additional relationships known to exist between the variables.

Now clearly Staicu's approach gives the appearance of providing a greater information input for, in addition to being given the list of variables and their dimensions, we are also told:

1. that only 1 DP enters the result;

2. that the exponents of the variables are in some sense minimal;

3. the sign of each exponent.

This additional information may not, however, be completely reliable and it is therefore difficult to quantify its effect upon the precision of the result without resort to probability theory. We may say that although there is an apparent increase in precision to unity, in that we conclude the analysis with one DP only, this result may, nevertheless, not be wholly reliable.

● 9.5   The expansion of a function as a series

The expansion in the form of a series of an undetermined function ϕ may sometimes help in resolving it. We give a few examples of this approach below.

● 9.5.1   A further approach to Poiseuille's problem

Consider the velocity of the fluid in steady laminar flow at a point at distance a from the axis of a circular tube of radius r. Analysis of this problem in the MLT system gives

$$v = \frac{dp}{dx} \cdot \frac{1}{\mu} \cdot r^2 \, \phi\left(\frac{a}{r}\right) \tag{1}$$

We require now to resolve the undetermined function ϕ, and argue as follows.

Think of the tube radius as increasing from r to $r + dr$. The fluid velocity at $a = r$ will then increase from 0 to dv. In this new situation the fluid flow for $a \leqslant r$ will be effectively contained within a (fluid) tube moving with a velocity dv in the direction of flow; or, to rephrase this, the fluid for all points $a < r$ will be contained within a tube of radius r which is itself moving forward with a velocity dv. It follows immediately from the principle of superposition that for each value of $a \leqslant r$ the flow velocity will be uniformly increased by dv. This implies that $\partial v / \partial r$ (for constant a) will not be dependent upon a.

With this behind us we proceed to expand ϕ as a power series:

$$\phi = \alpha + \beta\left(\frac{a}{r}\right) + \gamma\left(\frac{a}{r}\right)^2 + \partial\left(\frac{a}{r}\right)^3 + \ldots$$

Substituting in equation 1 and differentiating with respect to r gives for constant a

$$\frac{\partial v}{\partial r} = \frac{dp}{dx} \cdot \frac{1}{\mu} \cdot \left[2\alpha r + \beta a + 0 - \partial \frac{a^3}{r} - 2\epsilon \frac{a^4}{r^2} - 3\kappa \frac{a^5}{r^3} - \dots \right]$$

But we have shown that $\partial v/\partial r$ is independent of a, and it follows that $\beta = \partial = \epsilon = \kappa$ $\dots = 0$.

Reverting to the original equation entailing a fixed r, we see that the condition of non-slip at the tube boundary gives $\phi(a/r) = 0$ at $a = r$, entailing $\alpha + \beta + \gamma + \dots = 0$.

These two restrictions on the coefficients of the series give $\alpha = -\gamma$, all other coefficients being zero. Equation 1 may then be rewritten:

$$v = \frac{dp}{dx} \cdot \frac{1}{\mu} \cdot r^2 \left(\alpha - \alpha \frac{a^2}{r^2} \right)$$

$$= k \cdot \frac{dp}{dx} \cdot \frac{1}{\mu} \cdot r^2 \left(1 - \frac{a^2}{r^2} \right)$$

which is the required solution and which may be confirmed by an examination of the basic Navier–Stokes equations.

Note that in the foregoing argument we are justified in expanding ϕ in terms only of positive powers of (a/r), for the existence of any negative powers would involve infinite velocity at $a = 0$.●

●*9.5.2   Cooling of a tunnel wall*

The expansion of a function needs considerable care as difficult questions of convergence are liable to arise. As an example of the subtleties that may be encountered, we discuss the cooling of the periphery of a tunnel driven through a rock mass and ventilated by an air stream at a temperature lower than that of the rock. Putting θ_r as the original rock temperature, θ_a as the air temperature and θ as the (variable) temperature of the tunnel periphery at time t, we have the following table:

Physical quantity	Symbol	M	L	T	Θ
Fall of peripheral temperature below initial value	$\theta_r - \theta$	0	0	0	1
Excess of peripheral temperature over air temperature	$\theta - \theta_a$	0	0	0	1
Time	t	0	0	1	0
Thermal conductivity of rock	κ	1	1	−3	−1
Thermal capacity of rock per unit volume	h_v	1	−1	−2	−1
Heat-transfer coefficient (rock to air)	ϵ	1	0	−3	−1
Radius of tunnel	r	0	1	0	0

In commenting upon this, the thermal capacity of the rock is considered in terms of unit volume, thus avoiding the necessity for treating rock density as an additional variable. The coefficient of heat transfer, ϵ, is defined as the quantity of heat passing from rock to air per unit time per unit area of exposed rock per unit temperature difference. The quantity of air passing through the tunnel and its thermal capacity are not considered as significant variables since we are preferring to operate directly with the temperature difference between rock and air, which will itself be a function of these two omitted quantities.

Analysis now yields the equation:

$$\frac{\theta_r - \theta}{\theta - \theta_a} = \phi\left(\frac{\kappa t}{h_v r^2} , \frac{er}{\kappa}\right) \tag{1}$$

which may be expanded as a Maclaurin series by arguing as follows. For small decreases in the peripheral temperature we may write $(\theta_r - \theta) = \Delta\theta$ and, since the variable t represents the time taken for the temperature to fall from θ_r to θ, we may also write Δt for t. Equation 1 then takes the special form

$$\frac{-\Delta\theta}{\theta - \theta_a} = \phi\left(\frac{\kappa \Delta t}{h_v r^2}, \frac{er}{\kappa}\right)$$

which may be expanded to give

$$\frac{\Delta\theta}{\theta - \theta_a} = \phi\left(0, \frac{er}{\kappa}\right) + \frac{\kappa \Delta t}{h_v r^2}\, \phi'\left(0, \frac{er}{\kappa}\right) + \frac{1}{2!}\left[\frac{\kappa \Delta t}{h_v r^2}\right]^2 \phi''\left(0, \frac{er}{\kappa}\right) + \ldots$$

Since $\Delta\theta = 0$ at $t = 0$, the first term of this series is zero. If we may now neglect terms involving ϕ'' and higher derivatives, it follows that

$$\frac{d\theta}{\theta - \theta_a} = A \cdot \frac{\kappa\, dt}{h_v r^2} \tag{2}$$

where A is a function of er/κ. Integration gives

$$\ln(\theta - \theta_a) = -A \cdot \frac{\kappa t}{h_v r^2} + I$$

The constant of integration I is chosen to satisfy the condition that $\theta = \theta_r$ at $t = 0$ and this leads to

$$\theta = (\theta_r - \theta_a)\, e^{-A\kappa t/h_v r^2} + \theta_a \tag{3}$$

a solution which shows that the temperature of the rock falls away exponentially with time; this is valid, at least, in the initial stages of cooling.

In this treatment the question of convergency has been glossed over and we failed to consider the effect of the omission of the higher derivatives of the

Maclaurin series. We now note that the 'remainder' involving these terms is

$$R = \frac{1}{2!} \left[\frac{\kappa(\Delta t)}{h_V r^2} \right]^2 \phi'' \left(\beta \frac{\kappa \Delta t}{h_V r^2} , \frac{er}{\kappa} \right)$$

where β takes some value such that $0 < \beta < 1$. In considering this remainder we see that it is certainly true that $R \to 0$ as, and indeed more rapidly than, $t \to 0$, but we have no justification for taking Δt as small except in the early stages of cooling. Despite this objection, however, the solution given by equation 3 does, in fact, fit observed cooling rates rather closely over quite prolonged periods. ●

● *9.5.3   Underwater explosions*

The difficulties of this approach are nevertheless formidable and will be illustrated by a further example. Consider the maximum pressure p developed in a shock wave generated by an underwater explosion at distance r from the observer. This pressure will depend upon the initial gas pressure p_0, which, in turn, depends upon the nature of the explosive used. Other significant variables will be the mass of explosive detonated m, the density of water ρ, and the bulk modulus of the water K. Dimensional analysis provides as alternative solutions:

$$\frac{p_0 - p}{p_0} = \phi \left(\frac{p_0}{K} , \frac{\rho^{1/3} r}{m^{1/3}} \right) \tag{1a}$$

$$\frac{p_0 - p}{p} = \phi \left(\frac{p_0}{K} , \frac{\rho^{1/3} r}{m^{1/3}} \right) \tag{1b}$$

Taking the first of these solutions, we put $p_0 - p = \Delta p$. Since where $r = 0$ we have $\Delta p = 0$, we may consider the value of r corresponding with Δp to be Δr. This gives

$$\frac{\Delta p}{p_0} = \phi \left(\frac{p_0}{K} , \frac{\rho^{1/3}}{m^{1/3}} \Delta r \right)$$

$$= \phi \left(\frac{p_0}{K}, 0 \right) + \frac{\rho^{1/3}}{m^{1/3}} \Delta r \cdot \phi' \left(\frac{p_0}{K}, 0 \right) + \dots$$

Now since $\Delta p = 0$ when $\Delta r = 0$, it follows that $\phi(p_0/K, 0) = 0$ and, in consequence;

$$\frac{\Delta p}{p_0} \approx \frac{\rho^{1/3}}{m^{1/3}} \cdot A \cdot \Delta r$$

where A is a function of p_0/K only.

Integrating and remembering that Δp must be a decreasing function of r gives as our solution

$$p = p_0 \left(1 - A \cdot \frac{\rho^{1/3}}{m^{1/3}} \cdot r \right) \tag{2a}$$

On the other hand, if we work with the alternative equation 1b we find that

$$\frac{\Delta p}{p} = \phi\left(\frac{p_0}{K}, \frac{\rho^{1/3}}{m^{1/3}} r \right)$$

which leads to

$$p = p_0 e^{-(A\rho^{1/3}/m^{1/3})r} \tag{2b}$$

where A is not necessarily the same function of p_0/K as was considered in equation 2a.

There is an obvious contradiction here. It is nevertheless interesting to observe that a partial reconciliation between the two results may be made by expanding equation 2b as an exponential series. The first two terms of this expansion will be equivalent to the expression given in equation 2a. For small values of r, then, the two results agree, but for larger values they diverge appreciably. This confirms what had previously been suspected, and it would seem that the use of expansions as an aid to dimensional analysis will often be valid over only a small range of the dependent variable. Where this becomes large, discrepancies due to convergency difficulties are liable to arise. •

9.6   Miscellaneous examples

In discussing the applications of the techniques of dimensional analysis, it is often difficult to draw a rigid line between dimensional and conventional physical arguments. Indeed, there is a long history of dimensional analysis being used as an aid to calculation rather than being regarded as a sacrosanct technique to be used independently of other methods. Sedov[66], for example, while relying basically upon dimensional methods, allows his arguments to encompass the widest variety of other techniques. He is thus able fully to determine the nature of the functions with which he is working, whereas, had he confined himself solely to dimensional methods, these would inevitably have remained indeterminate. In our own text we have not hesitated to use extraneous arguments in order to obtain an explicit solution in such problems as those of **6.4.3**, **7.4.5** and **7.5.1**.

We now set out two further problems illustrating this 'mixed' approach:

• *9.6.1   Incipient motion of particles on a river bed*

We consider the problem of determining the conditions under which the motion of sand grains at the bed of a river is initiated. In accordance with the contemporary approach, we seek to derive an expression for the critical shear stress at the river bed necessary to produce incipient motion, since this quantity is, perhaps, more directly of interest than is, say, the critical flow velocity. If we ignore the statistical aspect of the problem and make the simplifying assumptions of a uniform grain size and shape and a level bed, then we may list the significant variables as the fluid density ρ, the sand density ρ_s, the gravitational acceleration g, the characteristic grain diameter d, the fluid viscosity μ and, finally, the critical shear stress itself τ_c. The problem reduces to the determination of a solution to the equation:

$$f(\tau_c, \rho, \rho_s, g, d, \mu) = 0 \tag{1}$$

A direct approach yields three DPs, but a limited insight allows us to reduce the number of variables before attempting analysis. We observe that the forces and moments tending to initiate grain movement will depend primarily upon flow conditions, whereas those tending to oppose such movement will be proportional to the apparent weight of the grain. It follows that ρ_s and g have significance only in determining this apparent weight and they must, therefore, appear in the final solution in the form $(\rho_s - \rho)g$.

Using this as a single variable, we are able to replace ρ_s and g in equation 1 and, in consequence, we obtain a more informative solution involving 2 DPs only:

$$f\left[\frac{\tau_c}{(\rho_s - \rho)gd}, \frac{\tau_c^{1/2}d}{\rho^{1/2}\mu}\right] = 0$$

In accordance with convention this may be rewritten in terms of the immersed relative density of the grains, $\gamma = (\rho_s - \rho)/\rho$ and the critical friction velocity $u^* = (\tau_c/\rho)^{1/2}$. We have then

$$f\left(\frac{u^{*2}}{\gamma gd}, \frac{u^*d}{v}\right) = 0 \tag{2}$$

where the second DP will be recognised as a Reynolds number Re^*.

The nature of this function will, of course, depend upon the characteristic grain shape and upon the validity of the assumptions we have made; but its general form may, to some extent, be revealed by a further appeal to physical insight. Experience suggests that, for turbulent conditions, viscosity will not influence the problem directly, but only in so far as to ensure that the Reynolds number is within the turbulent range. It follows from this remark that for high values of (u^*d/v) we may expect that

$$\frac{u^{*2}}{\gamma gd} = \text{constant}$$

For lower values of the Reynolds number, where the grains will be completely covered by a laminar sublayer, the critical shear stress will not be influenced by the precise nature of the bed roughness and this will, then, be independent of d. It follows that, within this range, the function in equation 2 can only be of the form

$$f\left(\frac{u^{*2}}{\gamma g d} \times \frac{u^*d}{v}\right) = 0$$

and we have, therefore,

$$\frac{u^{*2}}{\gamma g d} = k \cdot Re^{*-1}$$

Thus the indeterminate nature of equation 2 has now been fully reduced for certain flow conditions. •

• *9.6.2   Plume behaviour*

Use may be made of dimensional analysis in the investigation of jet and plume behaviour. To provide a final example of the manner in which direct dimensional analysis is supplemented by physical reasoning, we consider the emission of a hot gas from a small-diameter chimney stack and investigate the associated buoyancy plume – that is, we treat the case in which the momentum of the emittant is not significant.

Without as yet specifying a particular problem, we list the pertinent variables associated with the situation (reference may be made to figure 19):

Figure 19

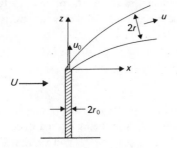

g	gravitational acceleration
$\dot{m}$	mass flow rate of emittant
r	radius of plume at specified section
r_0	stack radius
t	time
u	velocity of plume at specified section
u_0	velocity of gases at emission
U	characteristic cross-wind velocity

x horizontal distance from source
z vertical distance from source
ρ density of air
ρ_0 density of gases at emission
σ concentration of gases at specified section.

Plume phenomena are primarily due to density differences which are, in turn, due to temperature effects, but we shall consider densities as independent variables and shall not concern ourselves with their dependence upon temperature. We note, nevertheless, that there are as many as 12 variables which we have listed as significant, although, of course, not all of these will be included in any specific problem. As a further preliminary we describe the atmosphere as 'neutral' when the density/ temperature variation with height is such as to produce no additional buoyancy force on a parcel of emittant gas displaced slightly upwards, and as 'stable' when this variation results in a restoring force that counteracts the displacement.

Before proceeding with the dimensional analysis proper, we effect a reduction in the number of variables by introducing a physical argument: the buoyancy force on a parcel of gas emitted in time Δt will be $(\rho - \rho_0)g\pi r_0{}^2 \Delta z$. u_0 and a group characterising this buoyancy effect may then be $(\rho - \rho_0)g r_0{}^2.u_0$ ($\equiv MLT^{-3}$). As the dimension M occurs relatively infrequently in the list of variables, it is convenient to eliminate this from the 'buoyancy group' at this stage and we have in its place the buoyancy flux F defined as $F = (\rho - \rho_0/\rho). g r_0{}^2 . u_0 \equiv L^4 T^{-3}$.

We now take the case of a neutral atmosphere with no wind and derive expressions for the radius, velocity and concentration of the plume as a function of the height z. We write in turn each of the variables r, u and σ as functions of $\dot{m}, F$ and z, with dimensional analysis giving very simply

$$\left.\begin{array}{l} r = k \cdot z \\ u = k \cdot F^{1/3}/z^{1/3} \\ \sigma = k \cdot m/F^{1/3}z^{5/3} \end{array}\right\}$$

The next step is to introduce a cross wind and to ask how the plume rise varies with time or horizontal distance, these last two quantities being related by $x = Ut$. Here it is clear that z will depend only on F, U and t (or on F, U and x), that is

$$z = \phi (F, U, t)$$

or $z = \phi (F, U, x)$

These lead only to the incomplete solutions

$$z/Ut = \phi (F/U^4 t)$$

and $z/x = \phi (F/U^3 x)$

We are unable to resolve these results even if we assume a PP and introduce various simple physical conditions such as a monotonic increase in z with x. It has, however, been argued that F/U has significance as a stack parameter, since the component of F in the axial direction of a curving plume will be proportional to U. Granted this assumption we then have

$$z = \phi \, (F/U, t)$$

leading to

$$z = k \, . \, F^{1/3} t^{2/3} / U^{1/3} \tag{1}$$

or, alternatively, using $x = Ut$,

$$z = k \, . \, F^{1/3} x^{2/3} / U \tag{2}$$

We note, in passing, that Staicu's approach (9.4) fails to reveal these two results.

The reader may feel at this point that a degree of oversimplification and even of 'hindsight' has been introduced into our derivation. Admittedly, a more realistic consideration of the problem would have to take into account the effects of turbulence and would warrant, for instance, the inclusion in the problem of the friction velocity u^*. We will not ourselves be pursuing this approach, but the reader may be referred to Turner[69] for a detailed account of this and certain other points discussed.

In a stable atmosphere, any analysis of the plume configuration must take into account the fact that a maximum height is reached and that a further variable, characterising the degree of stability of the atmosphere, will be required. For an element of gas of mass $\rho \partial x \partial y \partial z$, a vertical displacement of ∂z will result in a restoring 'force' of $-\partial \rho . \partial x \partial y \partial z . g$. If we draw an analogy between the behaviour of a unit mass of gas and a simple spring, we are able to quantify this 'degree of stability' by writing the restoring force on the gas as

$$s = - \, \frac{\partial \rho \, . \, \partial x \partial y \partial z \, . \, g}{\rho \, . \, \partial x \partial y \partial z \, . \, \partial z}$$

or, in the limit, $s = - g.d\rho/\rho dz$ (per unit mass and displacement). This 'force' then, is analogous to a spring constant and has the dimensions of T^{-2}.

We now derive a series of results with a minimum of effort.

With no wind, the maximum height of plume rise may be written

$$z_m = \phi \, (F, s)$$

from which

$$z_m = k \, . \, F^{1/4} s^{-3/8}$$

The time taken for the gas to reach this maximum height will be

$$t_m = \phi \, (F, s)$$

from which

$$t_m = k \cdot s^{-1/2}$$

With a cross wind and using the parameter F/U in place of F we have

$$z_m = \phi \, (F/U, s)$$

from which

$$z_m = k \cdot (F/Us)^{1/3}$$

with t_m continuing to be given as above by $t_m = k \cdot s^{-1/2}$. ●

10

Model Laws and Similarity

10.1 Introduction

One of the more important applications of dimensional analysis lies in the possibility of predicting full-scale behaviour from observations made in the laboratory upon a suitable model. This approach, which can result in considerable economies and the avoidance of disastrous mistakes, has applications in a wide variety of fields. Model analysis may, for example, be used to investigate the stresses in a dam, the dynamic response of an aircraft wing, the deposit of silt in a harbour, the acoustic characteristics of a concert hall or the stability of an excavation profile in a mine.

The principles involved are of no great theoretical difficulty. For illustrative purposes we concern ourselves initially with the simplest of examples involving small oscillations of a pendulum. Dimensional analysis of this problem yields

$$f\,(t^2 g/l) = 0$$

entailing that $(t^2 g/l)$ is constant. Using subscripts m and p to denote variables pertaining to the model and the prototype respectively, we have as a necessary and sufficient condition for equivalence between the two systems:

$$(t^2 g/l)_m = (t^2 g/l)_p$$

Since g will remain unchanged in the two cases, this implies that

$$t_m{}^2/l_m = t_p{}^2/l_p$$

It follows that, by observing the period and length of the model pendulum, we may readily deduce the period t_p corresponding to a larger prototype pendulum of any required length l_p.

10.2 Physical similarity

If meaningful interpretations are to be made between model and prototype, it is necessary to be clear about the nature of the similarity that exists between the one and the other. Further, it will be advisable to examine the relationship that exists between similarity theory and dimensional analysis.

Similarity between systems may occur in a number of different ways: a musical score has similarity with the corresponding symphony; a map is similar to the region it represents; a model ship is similar to its prototype and a circuit involving an inductance, capacitance and resistance may be similar to a spring–mass–dashpot system in that they may both be described in terms of identical mathematical equations. There is, then, a wide range of meanings which may be attached to the word 'similarity'. Here we are concerned with 'physical similarity' and it becomes necessary to define this concept with care.

As a preliminary, we introduce the idea of 'homologous systems'. We say that two systems are homologous provided that each point and time of one (x_1, y_1, z_1, t_1) may be associated with a unique point and time of the other (x_2, y_2, z_2, t_2) and provided that a continuity criterion holds in that two neighbouring points on the one system correspond with two neighbouring points on the other. To phrase this differently, we require there to be a biunique, continuous correspondence between the points. Any two such systems and any two corresponding points contained within those systems are, then, said to be homologous.

'Physical similarity' is now defined to exist between two homologous systems when there is a biunique, continuous correspondence between the magnitudes of physical quantities associated with the homologous points of those systems. There is said to be 'complete' physical similarity if this applies to all physical quantities and 'partial' physical similarity if this applies to only some physical quantities. If the similarity is linear, as it is usually assumed to be, then the ratio of the magnitudes of a pair of 'homologous quantities', that is a pair of quantities associated with homologous points, will be the same, irrespective of which pair of points is considered.

Particular types of partial physical similarity may be distinguished:

Similarity involving L

Geometric similarity exists when the ratio of homologous lengths is constant, regardless of the directions considered.

Affine similarity exists when the ratio of homologous lengths in any particular direction is constant, but where this constant may vary with the direction considered.

Similarity involving M

Material similarity exists when the ratio of homologous point masses is constant.

Similarity involving T

Temporal similarity exists when the ratio of homologous durations is constant.

Similarity involving other reference dimensions may be defined in much the same way and it follows that we may discuss, say, 'electrical similarity', the constant ratio of homologous point charges, or 'thermal similarity', the constant ratio of homologous absolute temperatures.

A logical development of this approach leads to the consideration of similarity involving derived quantities. In particular:

Kinematic similarity exists when model and prototype are similar with regard to L and T.

This last is a sufficient condition for the ratio of homologous velocities to be constant. It is not, however, a necessary condition and some authors prefer to define kinematic similarity in terms of the necessary constancy of the ratio of homologous velocities.

Dynamic similarity exists when model and prototype are similar with regard to M, L and T.

This again is a sufficient condition for the ratio of homologous forces to be constant, and some authors prefer to define dynamic similarity in terms of the necessary constancy of the ratio of homologous forces.

In much the same way, other types of similarity may be defined and a variety of such concepts as 'aerodynamic' or 'magnetic' similarity are discussed in the literature. The point we make here is that all such types involve and are equivalent to the relevant combination of similarities based on the reference dimensions. In particular, where there is geometrical, material and temporal similarity, then there will be similarity for all mechanical quantities which may be represented in the *MLT* system. Shear stresses, for instance, or surface tensions would be similar and we would, in consequence, have 'shear-stress similarity' or 'surface-tension similarity'.

Note, in passing, that similarity with respect to a physical quantity may be alternatively defined in terms of *differences*. In fact we find that:

1. if in one system the difference in the values of a quantity measured at a pair of points x and y is D, and

2. if in a second system the difference in values of that quantity measured at a pair of points homologous to x and y is $k.D$, and

3. if the ratio k is constant between all sets of homologous point pairs,

then the two systems are similar with respect to the quantity concerned.

It is left as an exercise to the reader to show that this definition is equivalent to that made in terms of a constant ratio between absolute values.

Let us examine in more detail the conditions for dynamic similarity. It will be clear that wholly different kinds of force may simultaneously be acting within the same situation and that these may appear in a wide variety of forms. They may, for instance, be gravitational, inertial, magnetic etc. Now if full dynamic similarity is to be preserved, it will be necessary that each of these types of force, where it exists, must be separately similar—that is, the same constant of proportionality must apply between the two systems with regard to each and all of the forces present.

We illustrate this by examining a situation in which only two forces, say viscous and inertial, are significant. It is easy to see that, in such circumstances, the similarity criterion reduces to the preservation of a DP, that is (f_1/f_2). If, then, full dynamic similarity is to be present, both the viscous and the inertial forces acting at a point in the one system must be in constant ratio to the corresponding forces acting at the homologous point in the other system.

More specifically, consider an element of volume $\partial x . \partial y . \partial z$. The viscous force will then be given by, say, $\mu(\partial u/\partial y).\partial x.\partial z$ while the inertial force will be $\rho.\partial x.\partial y.\partial z.u.\partial u/\partial x$. Note here that u is the elemental velocity and that $u.\partial u/\partial x$ has the dimensions of acceleration. It follows that the relationship between the viscous forces of model and prototype will be

$$\left(\mu\frac{\partial u}{\partial y} . \partial x . \partial z\right)_{\mathrm{m}} = k_F . \left(\mu\frac{\partial u}{\partial y} . \partial x . \partial z\right)_{\mathrm{p}} \tag{1}$$

where k_F is the relevant proportionality constant. Similarly, for the inertial forces we have

$$\left(\rho . \partial x . \partial y . \partial z . u\frac{\partial u}{\partial x}\right)_{\mathrm{m}} = k_F . \left(\rho . \partial x . \partial y . \partial z . u\frac{\partial u}{\partial x}\right)_{\mathrm{p}} \tag{2}$$

Dividing equation 2 by equation 1 gives

$$\left(\frac{\rho u \partial y^2}{\mu . \partial x}\right)_{\mathrm{m}} = \left(\frac{\rho u \partial y^2}{\mu . \partial x}\right)_{\mathrm{p}}$$

Since similarity in L and T exists, the ratio of any pair of corresponding lengths will be the same and hence we see that $(\partial y^2/\partial x)_{\mathrm{m}}/(\partial y^2/\partial x)_{\mathrm{p}}$ may be replaced by $l_{\mathrm{m}}/l_{\mathrm{p}}$ where l is some characteristic length appearing in the situation considered, say the diameter of a duct or the radius of a projectile. Again the ratio of elemental velocities $u_{\mathrm{m}}/u_{\mathrm{p}}$ may be replaced by $U_{\mathrm{m}}/U_{\mathrm{p}}$ where U is a characteristic velocity, say the velocity along the axis of a tube. Making these substitutions we have

$$(\rho Ul/\mu)_{\mathrm{m}} = (\rho Ul/\mu)_{\mathrm{p}}$$

The DP we have obtained is, of course, the Reynolds number and the required similarity criterion therefore entails that the Reynolds number must remain constant between model and prototype — a result which may also be obtained as a straightforward exercise in dimensional analysis.

This last observation gives rise to the question of how far similarity theory and dimensional analysis overlap and how far they differ. In considering this, it is important to appreciate that similarity theory has applications only to similar systems, say model and prototype, whereas dimensional analysis has a wider range of application, model laws being only one particular instance of its usage.

Similarity theory does not merely list the DPs, the values of which are to be preserved, in any random fashion. It selects them, rather, in a meaningful way. Thus a valuable physical insight into the meaning of Reynolds number has been gained from the example just considered and we have been able to interpret it as the ratio of inertial to viscous forces within a fluid. The magnitude of the Reynolds number serves, then, to indicate the relative effects of these two forces in the situation under analysis. In much the same way, other requirements of similarity may be interpreted as a requirement of the constancy of some other DP between model and prototype.

Information concerning the appropriate DP which is to be held constant is obtained only as a result of a detailed study of the problem. Dimensional analysis will certainly provide a set of DPs based on considerations involving the significant quantities involved, but it does not concern itself with the meaning and the interpretation of the DPs that happen to be thrown up. It follows that the same results may be obtained by resorting either to dimensional analysis or to similarity theory, both approaches having their own advantages.

These remarks pose the question of what prerequisite must exist before dimensional analysis may be successfully applied to a model problem. This prerequisite is not that complete physical similarity must exist between model and prototype, for if physical similarity is given then the problem has already been solved and dimensional analysis becomes redundant. In order to throw light upon this it will be instructive to turn our attention to the following set of dependent variables:

1. the period of small oscillations of a pendulum,

2. the period of a deepwater surface wave,

3. the time of free gravitational fall through a given distance.

In all three cases we have

$$t = \phi(l, g)$$

where t is the appropriate period of time; l is the length of pendulum, length of wave, or distance of fall; g is the gravitational acceleration. In each case, dimensional analysis yields

$$t = k \cdot (l/g)^{1/2}$$

and we might, in consequence, be tempted to use, for example, a pendulum to represent a model of a deep-water wave. The constant k, however, takes a different value in each of the three cases, being equal respectively to 2π, $(2\pi)^{1/2}$ and $2^{1/2}$; and so, although essentially the same variables are involved, we must, from the point of view of similarity theory, treat each problem separately. It would seem, in consequence, that a sufficient condition for the application of dimensional analysis to a model problem is that model and prototype are homologous. We say here sufficient, rather than necessary, because there is no reason why a constant k should not assume identical values in two essentially dissimilar situations based upon the same set of variables. In such a case it would certainly be true that a model in one of the systems would provide information concerning a prototype in the other, even though no physical similarity existed.

10.3   The underlying principle of model laws

We consider more generally the principles underlying model laws. The dimensional analysis of any problem will result in an equation of the form

$$f(\pi_1, \pi_2, \pi_3, \ldots, \pi_n) = 0$$

which may be expressed explicitly for π_1 as

$$\pi_1 = \phi(\pi_2, \pi_3, \ldots, \pi_n)$$

Now if it be possible to allot certain values to $\pi_2, \pi_3, \ldots, \pi_n$, then the value of π_1 will be fully determined. (The case of multivalued functions will be considered shortly.) And so if model and prototype be constructed in such a way that

$$(\pi_j)_m = (\pi_j)_p \text{ for } j = 2, 3, \ldots, n$$

it will follow that

$$(\pi_1)_m = (\pi_1)_p$$

This result permits us to obtain information about an unknown variable within $(\pi_1)_p$ as a result of observations made upon the model. The maintainance of a constant DP does not, of course, necessarily involve keeping constant the quantities contained within that DP; indeed the fact that the length dimension varies between model and prototype generally entails a compensating variation in one or more of the other quantities.

   An example will make this clear. Let us examine, by means of a scale model, the problem of determining the stresses arising as a result of impact between two ships in collision. Analysis shows that the relevant DPs are those contained in the equation

$$f(\pi_1, \pi_2, \pi_3, \pi_4) = f\left(\frac{\sigma l^3}{mv^2}, \frac{El^3}{mv^2}, \nu, \frac{m}{\rho l^3}\right) = 0$$

s detailed below apply only to student textbooks.

ook which they are to use in their teaching and which on their recommendation
within the next twelve months.   Please write "recommended" overleaf against
return the information requested below.

....

....

....

....

o the book(s) you are keeping.

.... Date... No. of Students.........................

.... Date... No. of Students.........................

....

....

....

....

....

PT029417   E.8.

Dr. R.F. Cheeney
Grant Insitute of Geology
University of Edinburgh
Kings Buildings
West Mains Road
Edinburgh.

WHIC
28 DA
may b

PLEAS
full or

Edward Arnold (Publishers) Ltd.,
WOODLANDS PARK AVENUE, WOODLANDS PARK,
MAIDENHEAD, BERKS.   Tel : Littlewick Green 3104
VAT Reg. No.   207 9178 50

511:

ISBN	QTY.	AUTHOR & TITLE
0.7131.33473	1	✱ISAACSON. DIMENSIONAL
TOTAL NO. OF BOOKS	1	

N.B. BOOKS M.
MUST BE PAI
OR RETURN

BOOKS NOT INVOICED ABOVE V
WHEN AVAILABLE. APPROXIMA
SHOWN ABOVE/ON ATTACHED

where the stress is σ, the Poisson's ratio is ν, the velocity of impact is v and a characteristic length is l. We arrange our investigation in such a way as to maintain π_2, π_3 and π_4 constant between model and prototype and it follows that we obtain the information required concerning the stresses set up from the further relationship that is entailed, namely

$$(\pi_1)_m = (\pi_1)_p$$

If model and prototype are constructed of the same material, it will be ensured that E, ρ, ν and m/l^3 have the same values in each system. This means that $(\pi_3)_m = (\pi_3)_p$ and $(\pi_4)_m = (\pi_4)_p$. We may, moreover, further ensure that $(\pi_2)_m = (\pi_2)_p$ by maintaining the impact velocity v unchanged between model and prototype. As a direct result it will then follow that $(\pi_1)_m = (\pi_1)_p$ or

$$\frac{\sigma_m}{(m/l^3)_m (v^2)_m} = \frac{\sigma_p}{(m/l^3)_p (v^2)_p}$$

and, in consequence,

$$\sigma_m = \sigma_p$$

It follows that, provided we conduct our observations in accordance with requirements set out, the stresses set up in the prototype will be identical with those recorded in the model.

All this is straightforward enough. Let us, then, proceed to the case of multivalued functions and suppose that:

$$\pi_1 = \phi\,(\pi_2, \pi_3, \dots, \pi_n)$$

is multivalued for certain values of $\pi_2, \pi_3, \dots, \pi_n$. The foregoing argument will be unchanged provided that equivalent or corresponding values of π_1 are considered and so, if particular values of $(\pi_2, \pi_3, \dots, \pi_n)$ result in more than one value of π_1 — say $\lambda_1, \lambda_2, \dots,$ — then we shall have

$$(\lambda_1)_m = (\lambda_1)_p$$

$$(\lambda_2)_m = (\lambda_2)_p \quad \text{etc.}$$

We illustrate this with an example. Consider a two-degree-of-freedom, undamped system of springs and masses which is characterised by the equation

$$f\left(\frac{\omega}{(k_1/m_1)^{1/2}}, \frac{k_1}{k_2}, \frac{m_1}{m_2}\right) = 0$$

where k_1 and k_2 are the two spring constants, m_1 and m_2 are the two masses and ω is a frequency (see figure 20). Now if

$$\left(\frac{k_1}{k_2}\right)_m = \left(\frac{k_1}{k_2}\right)_p$$

and if

$$\left(\frac{m_1}{m_2}\right)_m = \left(\frac{m_1}{m_2}\right)_p$$

we then have

$$\left(\frac{\omega}{(k_1/m_1)^{1/2}}\right)_m = \left(\frac{\omega}{(k_1/m_1)^{1/2}}\right)_p$$

Figure 20

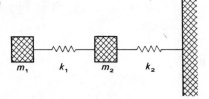

But there will be two values of π_1 corresponding to the two natural frequencies of vibration. However, the given relationship will remain valid provided always that the first solution (corresponding to, say, the lower frequency) is not equated to the second solution (corresponding to the higher frequency).

The next point we make is that it is often convenient to make a distinction between the dependent and the independent variables of a problem, despite the fact that all variables may strictly be 'interdependent'. In carrying out an analysis it is then usually arranged that the dependent variable should appear in one DP only, say π_1. The advantage of this approach is that, since the appearance of the variable whose value is of particular interest is confined in this way, all other DPs may be held unchanged between model and prototype (cf. 4.3, where we discussed the selection of arbitrary values for indices in order to restrict the manner in which variables appeared in the set of DPs).

Illustrating this point, we consider the problem of the underwater explosion (9.5.3). With the previously established notation, analysis gives

$$f(\pi_1, \pi_2, \pi_3) = f\left(\frac{p}{p_0}, \frac{p_0}{K}, \frac{\rho r^3}{m}\right) = 0$$

If we are interested in finding the maximum shock-wave pressure p developed at a particular distance r from the explosion, we shall write our equation as

$$\frac{p}{p_0} = \phi\left(\frac{p_0}{K}, \frac{\rho r^3}{m}\right)$$

thus ensuring that the dependent variable p is confined to its appearance in π_1. Assuming that the characteristics of water and explosive are unchanged between model and prototype, then ρ, p_0 and K will be identical in both systems, thus ensuring that $(\pi_2)_m = (\pi_2)_p$. Suppose next that the quantity of explosive used in the model is one thousandth of that to be used in the full-scale explosion; in order to maintain the value of π_3 in both systems, we must make

$$\left(\frac{\rho r^3}{m}\right)_m = \left(\frac{\rho r^3}{m}\right)_p$$

Thus

$$(r^3)m = \frac{m_m}{m_p}(r^3)_p = \frac{1}{1000}(r^3)_p$$

and

$$r_m = \frac{1}{10} \cdot r_p$$

Hence for a valid interpretation we must record the pressure at a distance from the centre of the model explosion equal to one tenth of the distance which will concern us when the prototype explosion takes place. Following this procedure we then argue that, since π_2 and π_3 have the same values in each system, so also must π_1. It follows from the fact that p_0 is a constant characteristic of the explosive used that $p_m = p_p$, which is the desired result.

If, however, we had been interested in r as the dependent variable and in determining, for instance, the distance at which the shock wave pressure attenuates to 1% of its initial magnitude, then we should have had to hold π_1 $(= 1/100)$ together with π_2 constant in both systems. The constancy of π_3 would then have been ensured and there would have been no difficulty in determining the required distance from appropriate observations upon the model.

10.4 Scale factors

An alternative method of regarding the relationship between model and prototype involves the use of scale factors. This does not represent a theoretically independent technique but consists rather of a convenient tool which the analyst should have available.

The scale factor of a variable x is defined as the ratio of the magnitude of x at a point in the model to the magnitude of the same variable at the homologous point in the prototype. We have, then, $k_x = x_m/x_p$. With this notation we may consider, for instance, a scale factor for length k_l, for force k_f or for Young's modulus k_E.

From our discussion on similarity, it follows that the scale factors have unique values for each of the variables they represent and, having decided upon one or more of these factors, we may then use known dimensional relationships to derive any other scale factors in which we may be interested. This procedure enables us to determine the conditions that underly the successful conduct of an investigation. Suppose that dimensional considerations applied to a certain situation involving a structural framework show that some DP, say (f/EI^2), is to be preserved between model and prototype. We have then to maintain the equality

$$\left(\frac{f}{EI^2}\right)_{\mathrm{m}} = \left(\frac{f}{EI^2}\right)_{\mathrm{p}}$$

Now k_l will be a factor prescribed by the scale chosen for the model and, provided the same material be used in both systems, we shall have:

$$k_l = l_{\mathrm{m}}/l_{\mathrm{p}}$$

$$k_E = E_{\mathrm{m}}/E_{\mathrm{p}} = 1$$

from which it follows that

$$\frac{f_{\mathrm{m}}}{f_{\mathrm{p}}} = \frac{E_{\mathrm{m}}}{E_{\mathrm{p}}} \cdot \frac{(l_{\mathrm{m}})^2}{(l_{\mathrm{p}})^2}$$

and $k_f = (k_l)^2$

showing that in this case the scale factor for force will be equal to the square of the scale factor for length.

Scale factors may also be regarded from a rather different viewpoint. If two systems have geometrical, material and temporal similarity (10.2), and if the L, M and T ratios between model and prototype are k_l, k_m and k_t, then the scale factor for any required quantity may be immediately determined from a consideration of the dimensional representation of that quantity. We may say at once, for instance, that the scale factor for acceleration will be $k_a = k_l/(k_t)^2$.

We give an example illustrating that the use of scale factors is not confined to dimensional analysis and that our use of them may be based directly upon a knowledge of the quantities involved or on the basic equation which relates to the situation. Thus, for a harmonically oscillating mass with linear damping, the equation of motion (for the prototype) will be

$$m_{\mathrm{p}} \frac{d^2 x_{\mathrm{p}}}{dt_{\mathrm{p}}^2} + \lambda_{\mathrm{p}} \frac{dx_{\mathrm{p}}}{dt_{\mathrm{p}}} + s_{\mathrm{p}} \cdot x_{\mathrm{p}} = 0$$

The relevant scale factors are $k_m = m_{\mathrm{m}}/m_{\mathrm{p}}, k_x = x_{\mathrm{m}}/x_{\mathrm{p}}, k_t = t_{\mathrm{m}}/t_{\mathrm{p}}, k_\lambda = \lambda_{\mathrm{m}}/\lambda_{\mathrm{p}}$ and $k_s = s_{\mathrm{m}}/s_{\mathrm{p}}$.

By making the appropriate substitutions in the variables of the prototype equation we have

$$m_m \frac{d^2 x_m}{dt_m^2} + \lambda_m \frac{k_m}{k_\lambda k_t} \cdot \frac{dx_m}{dt_m} + s_m \frac{k_m}{k_t^2 k_s} \; x_m = 0$$

But it will be clear that this equation will also apply to the model and that it may be alternatively represented by

$$m_m \frac{d^2 x_m}{dt_m^2} + \lambda_m \frac{dx_m}{dt_m} + s_m \cdot x_m = 0$$

A comparison between these last two equations shows that the 'model laws' relating to this particular problem will be

$$\frac{k_m}{k_t k_\lambda} = \frac{k_m}{k_t^2 k_s} = 1$$

or, simplifying, $k_\lambda = k_m/k_t$ and $k_s = k_m/k_t^2$.

As a further illustration of the possibility of deriving and using scale factors without recourse to dimensional analysis, we set out a neat example due to Jupp[39]. It is intended to build a pump capable of lifting 5 m^3/s against a head of 250 m and operating at a speed of 500 rev/min. A 1/10 scale model is constructed for testing and we have, therefore, as the scale factor for length $k_l = 1/10$. Provided that the model is pumping the same fluid as the prototype we shall have $k_\rho = 1$, that is, in terms of the scale factors for mass and length, $k_m/k_l^3 = 1$. It follows that $k_m = 1/1000$. Since gravity is assumed constant for each system, $k_g = 1$ and we have $k_l/k_t^2 = 1$, whence $k_t = k_l^{1/2} = 1/\sqrt{10}$. Now the speed ratio $k = k_l/k_t = 1/\sqrt{10}$ and the model should, therefore, run at $1/\sqrt{10} = 0.316$ of the speed of the prototype, that is at 158 rev/min. The scale factor for delivery will be $k_Q = k_l^3/k_t = \sqrt{10}/10^3 = 0.00316$ and the delivery of the model will, therefore, be 5 x 0.00316 = 0.0158 m^3/s against a head of 250 x k_l = 25 m. The power scale factor $k_p = k_m k_l^2/k_t^3$ is equal to 0.000316, and it follows that the power requirement for the prototype may be readily calculated in terms of that consumed by the model. This argument, of course, assumes equal efficiencies in a geometrically similar prototype and model and it will be noted that once again no explicit use of dimensional analysis is made.

10.5   Situations involving the strict inapplicability of model laws

In many cases it is difficult, impracticable or impossible to hold constant the values of DPs between model and prototype. This may be due to the fact that some quantities are confined to a restricted range of values or even to specific values, as is liable to be the case where material properties such as elastic moduli or thermal capacities are concerned, or where dimensional constants such as g are involved.

As an example of this difficulty, we make an investigation into the drag force on a ship. Dimensional analysis of the problem gives, in terms of standard dimensionless numbers,

$$C_D = \phi \, (Re, \, Fr)$$

If now we work with, say, a 1/20th scale model ($k_l = 1/20$), then the values of the other variables required in the model test may readily be determined by the methods we have established.

Equating the Froude number between model and prototype gives

$$\left(\frac{v^2}{lg}\right)_m = \left(\frac{v^2}{lg}\right)_p$$

Hence

$$\left(\frac{v_m}{v_p}\right)^2 = \frac{l_m}{l_p} \cdot \frac{g_m}{g_p}$$

$$= k_l \quad \text{(since } g \text{ remains unchanged)}$$

and so

$$k_v = k_l^{\,1/2}$$

We next equate the Reynolds number of the model to that of the prototype:

$$\left(\frac{vl}{\nu}\right)_m = \left(\frac{vl}{\nu}\right)_p \quad (\nu \text{ being the kinematic viscosity)}$$

giving

$$\frac{\nu_m}{\nu_p} = \frac{v_m}{v_p} \cdot \frac{l_m}{l_p}$$

$$= k_v \cdot k_l$$

or

$$k_v = k_l^{\,3/2}$$

For a 1/20th scale model, then, we have

$$k_v = \left(\frac{1}{20}\right)^{3/2} = 0 \cdot 011$$

which implies that v_m will have to be about $0 \cdot 011 \times 10^{-6}$ m^2/s (where 10^{-6} m^2/s is the approximate kinematic viscosity of water in the prototype at typical ambient temperatures).

It will be clear that it would be out of the question to use in the model a fluid which is to be characterised by so low a viscosity, and no wholly satisfactory way can be found of solving this technical difficulty. We are forced, then, to resort to approximate methods by allowing Re to differ between model and prototype and to make the assumption that changes in Re will result in relatively small effects upon the drag force within the range considered. Alternatively, we may fall back upon other methods in order to estimate the necessary correction that has to be made. The drag force may, for example, be considered as the sum of 'wave drag' (dependent upon the Froude number only) and 'viscous drag' (dependent upon the Reynolds number only). The latter is then estimated from boundary-layer theory or from a separate investigation that is initiated with this end in mind.

Complete similarity, then, although certainly a sufficient condition for the interpretation of model behaviour is not always a necessary one and, as we have seen, it may be difficult or impossible to impose. Consequently, deviations in the value of one or more DPs may be permitted to take place between the two systems, provided that care be taken that these deviations affect only DPs having a negligible influence upon the dependent variable or, alternatively, provided that the deviations affect the dependent variable in a known manner.

A further example of this same problem lies in the difficulty of obtaining a sufficiently high Reynolds number for tests on model aeroplanes. This is liable to happen when, for instance, the velocities required become too high for the available equipment to handle. Now if the effect of the higher and unobtainable Reynolds number is known to result in the transition from a laminar to a turbulent boundary layer, then this effect may sometimes be simulated by the use of a trip wire placed on the wing in such a fashion that the required turbulent boundary layer is established.

A major difficulty in the use of models is that variables which may be insignificant in the prototype are liable to become significant in the model. The effect of this is to extend the list of variables entering into the situation and hence the number of DPs proliferates, with added complication to the investigation. This is typical of the situation which occurs with the high velocities in a wind tunnel which, as we have seen, may be necessitated by the need to maintain a certain Reynolds number. These velocities will frequently be such that compressibility becomes an important quantity, even though it may be safely ignored in the prototype. Another illustration involving tunnels occurs when the walls of the tunnel are sufficiently close to the model to influence behaviour. There then arises the necessity for considering the additional DP (h/l), h being the wall separation and l a characteristic length of the model.

Again, the small linear dimensions of a model harbour may result in surface-tension effects becoming important, whereas these are always negligible in the prototype. And, if we are considering the flow of a particulate material through a silo, the effect of the grain size could be a significant variable in the model, even though it might be without influence in a full-scale silo. Any attempt to scale down the grain size in the model material could, moreover, lead to the introduction of extraneous effects such as agglomeration which would result in major difficulties of interpretation.

10.6   Distortion

In our consideration of geometric similarity, we stipulated that k_l must remain constant for homologous lengths. If a distinction is to be made between, say, vertical and horizontal lengths, in such a way that we have one scale factor k_x for horizontal lengths and a different scale factor k_z for vertical lengths, then the model is said to be distorted. If a further distinction be made between orthogonal horizontal lengths, then there will exist three unequal length scale factors $-k_x$, k_y and k_z — and we say that the model has a 'second degree' of distortion. In such cases, as will be seen from the definition in **10.2**, affine similarity will still exist.

Length distortion need not be the only type of distortion present. Where different processes taking place in the same system have different characteristic times, then time distortion may be introduced — that is there will exist scale factors k_{t_1} and k_{t_2}, these not necessarily being equal. Again, many other quantities such as density or force may often be advantageously distorted in certain types of model.

We are, then, led to ask under what conditions is it permissible to use distortion. It is frequently found convenient to distort a model even though it may not be possible to find a logical justification for the process and simply because experience has shown that the distortion is of such a nature as not to inhibit the provision of a reasonable solution. Or we may resort to distortion because we find that it may lead to a solution of certain aspects of a problem, despite the introduction of further complexities that have to be cleared up by supplementary arguments or observations. An example involving the successful use of the technique of distortion, even though it is not logically justified, is set out in **10.7**.

All distortion, however, does not fall within this pragmatic pattern, for under certain conditions it may be introduced with full assurance of its logical justification. This becomes possible where there exists a physical independence between two quantities, fundamental or derived. A distortion may then be effected by allowing different scale factors for each of these independent quantities. We recall here the remark made in **5.2**, where we discussed the possibility of making arbitrary changes in the scale of units where physical independence is known to exist. It will be profitable to consider illustrations of this important result.

In the elementary projectile problem mentioned in **5.3.2**, we noted that physical independence existed between horizontal and vertical lengths. In this type of situation, then, we may distort the x direction relative to the y direction and use a distorted model to represent a prototype. Suppose, for instance, that we attempt to represent the trajectory of a bullet by projecting a stone in such a fashion that it returns to the ground at a distance of 10 m from its starting point and after a flight time of 2 s. We notice that the highest point of the trajectory is 5 m above ground level. If now we distort horizontal distances by a factor of 500, we can immediately say that the bullet will strike the ground at a distance of 5000 m after a lapse of 2 s and, moreover, since vertical distances are left undistorted, the highest point of the trajectory will again be 5 m above ground level. Note that the horizontal component of velocity will suffer the same distortion as do horizontal lengths and the value of u_x, which is 5 m/s in the model, becomes 5 x 500 = 2500 m/s in the prototype.

A similar approach involving legitimate or logically justifiable distortion of physically independent length dimensions would enable us to predict, say, the behaviour of a prototype long, thin, bimetallic strip from a consideration of a short, thick model, since, as we saw in 5.3.5, the radial and axial length dimensions are here physically independent.

We re-emphasise this point, which does not appear to have been made explicit by other authors: that where physical dependence exists in the reference dimensions or in two or more of the derived quantities, then a corresponding distortion may safely and legitimately be introduced, provided always that this independence persists in both systems.

That this last proviso is necessary is readily shown to follow from the comments made in 7.3, where we mentioned that if a laminar boundary layer is assumed thin then it will be possible to treat the flow direction and the normal to this as independent. In consequence, it is legitimate to introduce a distortion and, for example, to use a boundary layer of one thickness to represent that of another. But if the distortion be carried too far, then the 'thin-boundary-layer assumption' is no longer valid and consequently there is no longer physical independence between the two directions.

It is, then, clearly necessary, if a distortion is to be introduced, that the physical independence upon which it is based shall hold both in model and prototype rather than in one system only. Subject to this condition, the correspondence will be such that valid interpretations may be made from the behaviour of the distorted model.

In a number of situations it may be necessary to assume an anisotropic material property in order to ensure physical independence between orthogonal directions. Young's modulus, for instance, may be treated as three independent variables E_x, E_y and E_z, and in order to achieve the conditions for a legitimate distortion it may be necessary to apply to each of these quantities a different scale factor. A case in point is the example (5.5.1) in which we investigated the lateral pressure beneath the earth's surface. Using Z as the dimension corresponding to the vertical direction and L as that corresponding to directions in the horizontal plane, we obtain the dimensionally homogeneous equation

$$P = \frac{v_z}{1 - v_l} \cdot \rho g z$$

where $v_z \equiv L^2 Z^{-2}$ and $v_l \equiv [1]$. It is therefore possible to use a model in which v_z is no longer held equal to v_l. The possibility of using anisotropy in distorted models, together with practical methods of achieving it, is discussed by Costa[11].

● 10.7   An example involving distortion

Problems in hydraulic engineering are of great economic importance and are frequently intractable when approached by theoretical methods. Extensive and successful use is made of models in this field, even though the distortion that has to be used is frequently not to be justified in terms of physical independence. For open-channel flow with fixed beds, a dimensional analysis indicates that the Froude and Reynolds numbers should be held unchanged between model and prototype if the flow is to be correctly simulated; but, as has already been shown (10.5), it is seldom practicable to preserve both these numbers. It follows that complete similarity cannot be achieved. There are, moreover, further difficulties associated with the substantially reduced lengths that are often necessary in these models. Consider the scaling down of a large river in order to work with a model confined to the bounds of a laboratory. Quite apart from the general inconvenience entailed by the use of the small lengths with which we have to work, the surface-tension effects, which are unimportant in the river itself, now assume major significance. Furthermore, the flow in the model will invariably be laminar and thus completely different in nature from the turbulent flow that characterises the prototype. Another problem lies in considering the roughness of the river bed. This is of great importance in that it affects the resistance to flow, but the roughness parameter will invariably be greater in the model than in the prototype and it follows that true geometric similarity cannot be achieved.

To see how these practical difficulties may be largely overcome, we examine the effects of introducing a vertical distortion with $k_z > k_x$, where z and x correspond respectively to the vertical and to any horizontal direction. As the depth is significant in determining both the state of flow and the effects of surface tension, the relatively greater depth used in the distorted model usually has the desired effect of ensuring that the flow remains turbulent and, further, that complications due to surface tension are minimised.

We are, however, still faced with the problem of roughness. It is thought that the roughness parameter, defined as the ratio of the mean height of roughness projections to the depth of flow, and the Reynolds number affect the problem only indirectly and by virtue of their contribution to the skin-friction coefficient, the position being, in fact, similar to that indicated in figure 16 (page 124).

In our distorted model, the physical considerations to be outlined below indicate that an increase in the skin-friction coefficient will be necessary. Defining the skin-friction coefficient as $C_f = \tau_w/(\frac{1}{2}\rho v^2)$, τ_w being the shear stress at the river bed, it can readily be shown that the condition for steady flow is simply:

$$C_f = 2\,Rgs/v^2$$

where s is the gradient and R the hydraulic radius – that is, the cross-section of the flow divided by the wetted perimeter.

Working in terms of scale factors we have: $k_g = 1$, $k_s = k_z/k_x$, $k_v = k_z^{1/2}$ (see **10.5**), while k_R will be dependent upon the cross-section profile. For the simpler case of a wide river, R approximates to the river depth and here, then, $k_R \approx k_z$. It follows that, since

$$k_{C_f} = \frac{k_R \cdot k_g \cdot k_s}{k_v^2}$$

we have

$$\frac{(C_f)_m}{(C_f)_p} = \frac{k_z}{k_x} > 1$$

It follows that the roughness parameter in the model must be so adjusted as to produce the appropriate skin-friction coefficient at the lower Reynolds number applicable to the model, and practical techniques for accomplishing this are in common use (see, for example, Yalin[72]).

Additional examples of interest may be drawn from the consideration of hydraulic models with moveable beds, that is beds consisting of sand or silt. Here further distortions, including independent 'hydraulic' and 'sedimentation' time scale factors, are used to advantage. Problems involving surface waves, however, are not solved by distortion, for the relatively greater depth of sea compared with river renders distortion less necessary.

In the final analysis, methods involving trial and error must inevitably be used in problems involving the types of complexity we are now considering. The model is made to represent phenomena that have already occurred and, if these results are satisfactory, then the procedure adopted may be extrapolated to take into account changes which are envisaged in the basic situation – often with gratifyingly successful results. •

10.8   Applications to biology

Lastly we mention a number of well-known scale effects apparent in the world of biology. These are dealt with on a systematic basis by Wentworth Thompson[68], but the general nature of such effects was appreciated as long ago as the seventeenth century when Galileo[27] wrote

> Clearly then, if one wishes to maintain in a great giant the same proportion of limb as that found in an ordinary man, he must either find a harder and stronger material for making the bones, or he must admit a diminution of strength in comparison with men of medium stature; for if his weight be increased inordinately he will fall and be crushed under his own weight. Whereas if the size of a body be diminished, the strength of that body is not diminished in the same proportion; indeed the smaller the body the greater the relative strength. Thus a small dog could probably carry on his back two or three dogs of his own size; but I believe that a horse could not carry even one of his own size.

Quite generally we notice that for animals of similar structure, weight will increase as the cube of the linear dimensions, that is, $W = k \cdot l^3$. But the cross-section of the leg increases only with the square of the linear dimensions and we have $A = k \cdot l^2$. It follows immediately that the compressive stress within the leg, that is $\sigma = W/A$, must vary as the first power of the characteristic length of the animal. Since there is clearly an upper limit to the compressive stress which may be supported by bone and tissue, there will be an upper limit to body size. In heavier animals, such as the elephant or hippopotamus, the cross-section of the leg does, in fact, have to be disproportionately large and this results in clumsiness of gait. The upper limit of magnitude may, then, by this means, be deferred — but it cannot be abolished. The argument does not, of course, apply to water-borne creatures, such as the whale, which may reach considerably greater sizes and weights than would be possible for creatures living on land.

Other arguments, approximate though they may be, indicate a variety of results applicable to structurally similar animals. For example:

1. The muscular force that can be exerted varies as the muscular cross-section or as the square of the characteristic length, that is $f = k \cdot l^2$. The work output will vary with the force exerted by the muscle multiplied by the extent of its contraction, which itself varies with the first power of the characteristic length. We thus have $W_1 = f \cdot s = k \cdot l^3$. Now if an animal can jump to a height h, the work expended will be equal to h multiplied by the weight of the animal, which in turn varies with the cube of its linear dimensions. The work required to effect the jump will then be $W_2 = k \cdot hl^3$. Since the capacity to perform work has already been shown to vary with l^3, it follows that the maximum height of jump will be largely independent of the size of the animal which is characterised by l. Taking the structure of the mouse, the dog and the elephant as approximately similar, the height that each is able to jump should be, and is, of about the same order of magnitude, despite a difference in their masses by a ratio of many thousands.

2. Small animals tend to be more efficient than large ones, as calculated on a power/weight basis, since the energy available for continuous output will vary with the surface of the lung, that is with l^2, whereas the weight varies with l^3. A small creature can move with ease an object many times its own weight, whereas a larger one will experience difficulty.

3. The vocal cords of small animals will be more highly pitched than those of large ones, since the frequency of vibration of vocal cords of similar density and tension varies inversely with l^2. A mouse squeaks; a lion roars.

4. The normal rate of locomotion tends to vary with $l^{1/2}$ since the advance per pace increases with l but the number of paces of unforced walking per unit time varies with $l^{1/2}$. This assumes that the swing of a limb does not depart too radically from that of a pendulum of similar density distribution.

5. A warm-blooded animal loses heat according to its surface area. Its intake of food varies with its stomach capacity, that is with its volume. Since the ratio of area to volume increases as the size of animal decreases, it follows that too small a mammal would have to spend its whole time eating in order to keep warm. In practice, certain shrews have to eat every 2 hours or so and mammals of a smaller size would seem to represent an impracticable proposition. (Pearsons[54] cites the case of a shrew weighing 3·6 g which ate food weighing 93 g in 8 days!)

6. Finally, among the protozoa we have such effects as an upper limit on size imposed by the fact that nourishment is absorbed through the cell or body surface which varies only as the square of the linear dimensions, whereas the food requirement will be related to the total weight which varies as the cube. Similarly, the use of cilia or flagella as a means of locomotion is possible only for very small organisms (see Sandon[62]). Another scale effect is that streamlining ceases to be advantageous among protozoa as, when in motion, they have only to contend with viscous forces rather than with the inertia of the surrounding medium. Moreover, with viscosity all-important, gliding passively through water becomes impossible. A small ciliated organism starts off at full speed as soon as the cilia become active and comes to an abrupt halt when the motive power of the cilia is cut off.

11

Miscellaneous Aspects of Dimensional Analysis

11.1 The necessary aspect of natural laws

There are those who hold that an implication of dimensional analysis is that physical laws are of a wholly necessary nature, being logically entailed by the structure of the quantities that are related; that physical laws are, in this sense, immanent; and that a degree of *a priori* knowledge of physical relationships is, in principle, possible.

Eddington[18] was influenced at least in part by considerations of dimensional analysis when he wrote

> I believe that the whole substance of fundamental hypotheses can be replaced by epistemological principles. Or, to put it equivalently, all the laws of nature that are usually classed as fundamental can be foreseen wholly from epistemological considerations.

We do not fully accept this point of view, for it is generally agreed that *a priori* knowledge is limited to fields of logic and pure mathematics. Contingent truths concerning the nature of the physical world can never be based entirely on *a priori* grounds, for we have first to establish principles by the application of an empirically based induction and only then can we make deductions from those principles.

In particular, we make the point that any physical argument, whether or not it be based on dimensional methods, must be rooted in the assumption that a phenomenon relates to the interaction of a certain restricted set of variables and no others. The validity of an argument rests firmly upon the truth of this assumption, and considerable experience of the nature of physical processes will usually be necessary before the selection of a significant set of quantities may be made.

There is, then, no shortcoming in the method of dimensional analysis when we are called upon to say that this quantity will be involved but not that, for it is clear that a similar decision as to what basic quantities enter into a situation must also be made when conventional analysis is undertaken. Indeed, as Duncan[15] has it,

> Whenever we frame a detailed mathematical investigation of a physical

phenomenon, we always have to begin by deciding what physical quantities are to be included in the investigation.

Dimensional analysis does not, and never should, pretend to constitute an exception to this rule.

Again, Bridgman[5] writes

In respect to this degree of dependence on past experience, dimensional analysis differs from no other scientific enterprise; for it is never possible to obtain factual information about any concrete physical situation by pure ratiocination.

But once the selection of well-defined quantities has been made, then it must follow, necessarily and inevitably, that those quantities can only be related in certain ways and subject to the major restrictions imposed by the requirements of dimensional homogeneity. *If* the period of a pendulum is to depend upon l and g, and upon l and g only, *then* we can say that it necessarily follows that $t \propto (l/g)^{1/2}$; but we cannot argue upon any *a priori* grounds that the period is in fact related to the two variables considered. Here physics remains firmly rooted in empiricism.

Having once made our selection of significant quantities, any subsequent result that may be deduced by the application of dimensional methods will be wholly necessary; indeed, it will be almost tautologous in its force. The tautology may be masked, but its effect will be much the same as when we divide m oranges among n boys and say there are m/n oranges per boy. It is, indeed, logically impossible that the dependent variable 'oranges per boy' could be represented by any other combination of the independent variables m and n.

We have purposely mentioned that the determination of a physical relationship is based upon the selection of the appropriate set of 'well-defined' quantities. We now make the additional point that the more information which can be embraced by the definitions, the greater the precision of the result that is potentially available. It is for this reason that a suitable extension to the set of independent reference quantities is likely to be useful, for definitions that can be based upon the extended set are liable to have a higher information content than are those based upon a restricted set.

We have at no time suggested that the selection of significant variables is necessarily a simple task. If we investigate the stress distribution at a distance r from the centre of a vertical shaft of circular cross-section and radius a, sunk in isotropic rock and subject to a hydrostatic pressure p, then a complete analysis shows that the principal stress in the radial direction is given by $\sigma_r = p(1 - a^2/r^2)$. Dimensional analysis would have yielded only $\sigma_r = p \cdot \phi(a/r)$. We could, however, have been forgiven had we assumed that the Young's modulus and the Poisson's ratio for the rock were also quantities which entered into the relationship. Analysis would then have yielded

$$\sigma_r = p \, . \, \phi \, (p/E, \, v, \, a/r)$$

which, without being in any way 'wrong', is certainly liable to mislead.

Again, in selecting the significant quantities, we may be doubtful whether or not to include molecular or atomic forces rather than macroscopic parameters such as viscosity or surface tension. It is clear, however, that the macroscopic quantities which occur in the elementary treatment of a problem represent a statistical average or an effective resolution of the more fundamental quantities that give rise to them. Accordingly we choose our quantities in such a way as is appropriate to the level at which we wish to carry out our investigation.

There is never any *a priori* procedure by which we may obtain information that a certain set of quantities each defined in a certain way, and that set alone, significantly affects the situation. Dimensional analysis pulls no rabbits out of hats. A man confined since birth in a cell and having no contact with the external world could learn nothing of that world so long as he had to rely upon ratiocination unaided by observation.

11.2   The possibility of associating an arbitrary DP with a physical relationship

There will, in general, be a linear relationship between $(n + 1)$ vectors of order n (see, for example, Littlewood[43]). It follows immediately that in the *MLT* system, or in any other system based on three reference quantities, any 4 variables selected at random may be put in the form of a DP. This, in itself, never implies that there is in fact a physical situation in which there occurs an actual relationship between the variables considered. We again are forced to rely upon our experience of the outside world to tell us whether the relationship we have deduced does or does not correspond to any known state of affairs. From the *a priori* viewpoint we can say only that *if* there is a relationship, *then* the relationship will be of this or that particular form.

There is an interesting approach to this question of whether or not there exists a physical relationship connecting a set of variables that happen to form a DP. This is based on the statement in 1.6 where we mentioned that numerical coefficients in physical equations tend to be small and neat. If, then, we deduce a DP and if we substitute observed values for the quantities which comprise it, we shall determine a number. Where that number is extremely large or small, we may take it as an indication that our DP corresponds to no physical relationship; where, however, the value of the DP is of the order of unity, then it may be, at least, worthwhile to investigate the matter further.

A classic, and frequently quoted, instance of this type of argument is due to Einstein[20], who, in the early days of the study of the specific heats of solids and their connection with quantum phenomena, suspected that there might be a relationship between the inter-atomic forces determining the elasticity of a solid and those concerned with the infra-red characteristic frequencies. He therefore considered the possibility of expressing compressibility in terms of the atomic mass; the distance apart of the atoms, represented by the number of atoms per unit volume; and a characteristic frequency. Analysis shows that there could well be such a relationship and that, if it existed, it would necessarily be of the form

$$\kappa n^2 N^{1/3} m = \text{constant}$$

where κ is the compressibility $((\partial V/\partial p)/V \equiv M^{-1}LT^2)$, m is the atomic mass, N the number of atoms per unit volume and n is the characteristic frequency.

Einstein proceeded next to consider copper as a special case and, substituting the known values of the quantities contained in the DP, found that the magnitude of the constant was indeed very close to unity. This was taken as compatible with the fact that the constant might well have arisen from the mathematical manipulation of certain unknown basic equations connecting the variables and he was, therefore, encouraged in his suspicion that a relationship did in fact exist. Final proof, however, was not obtained until the equations had actually been derived from basic principles by Debye, in his well-known analysis of specific-heat phenomena.

As a counter-example, Bridgman[5] supposes an investigation into a possible electrodynamic theory of gravitation. Within such a theory it might be suspected that a relationship exist connecting the gravitational constant with the charge on an electron, the mass of an electron and the velocity of light. Analysis shows that such a relationship could only take the form $G = k \cdot (e/m)^2$, with the velocity of light c dropping out. But as soon as we substitute the known values of G, e and m, we find that $k = 2.35 \times 10^{-43}$, which is of an impossibly small order of magnitude, and we reject the relationship as having a probability approaching zero. (See also problem 37 of Appendix 2.)

Such examples support our previous argument that *a priori* knowledge of natural laws, unsupported by observation, is not possible. We must not, however, be driven to the opposite extreme. The tendency to regard a law of nature merely as an observed regularity can be equally misleading. A physical law not only describes a phenomenon that is observed to take place; it is also entailed by the inner structure of that phenomenon and, once that structure (which includes the structure of the independent variables) is known, we see that the phenomenon could not be other than it is. If the quantities t, l and g and only those quantities are known to enter into a situation, then it is logically impossible that t should do other than vary with $l^{1/2}$. We are not, as Karl Pearson[53] suggested, restricted to asking 'how' in science — 'how' in the sense of 'in what manner'; we may also ask 'why' — 'why' in the sense of 'what is the underlying reason for', though not, of course, 'what is the purpose of'. Questions of purpose are best left to theologians.

11.3   Restrictions on the method of dimensional analysis

It will have become evident that the technique of dimensional analysis is such that it suffers from two severely limiting restrictions, namely

1. the numerical constant entering into a physical equation has to be left undetermined;

2. the nature of functions, where there is more than one DP, must also be left undetermined.

We ask why it is that these limitations should arise and whether it is possible that some refinement of the analysis might avoid them.

The position with regard to the undetermined constant is rather simple. We recall from **1.5** that an algebraic symbol in a physical equation stands not for a 'number' but for a 'number of specific units'. Now, dimensional analysis concerns itself wholly and essentially with the equivalence of the units: it never interests itself in the numbers associated with those units. It follows that no possible extension or refinement of dimensional analysis can lead, say, to the determination of the constant 2π in the equation $t = 2\pi(l/g)^{1/2}$.

There is, nevertheless, an approach based on dimensional analysis which leads to fully determinate results, even though these are liable to be of a largely trivial nature. To demonstrate this we have merely to differentiate the equation obtained as an exercise in dimensional analysis and then eliminate the undetermined constant between the original and the differential equation. Thus, starting from equation 2 of **5.6**, we have

$$\dot{m} = k \cdot \frac{\rho r^4}{\mu} \cdot \frac{dp}{dx}$$

$$\frac{\partial \dot{m}}{\partial r} = k \; 4 \cdot \frac{\rho r^3}{\mu} \cdot \frac{dp}{dx}$$

and thus

$$\frac{\partial \dot{m}}{\partial r} = 4 \frac{\dot{m}}{r}$$

which is wholly determinate, even though it may not be a result of any particular interest.

Similarly, despite the fact that we are not able to resolve an undetermined function by this method, we may be able to eliminate it entirely. Thus differentiating equation 1 of 2.5 we obtain the determinate (but essentially trivial) equation $dt/dl = t/2l$, which is valid for a simple pendulum no matter how great the amplitude of the (constant) angle of swing.

We now consider how far the restriction in the power of dimensional methods to determine functions is an essential characteristic and how far it may be an accidental limitation due to an imperfect development of the theory. It is clear that in many cases it can never be possible to extend the number of essential dimensions until we achieve the limiting position $(n - r) = 1$, for this would imply that the solution to each and every physical problem could be expressed as a PP, which is obviously impossible, for we know of many situations that may adequately be described only in terms of other types of function. It follows immediately that no refinement of analysis can be expected to yield complete solutions where this is the case.

It may occasionally be possible to obtain functions that appear superficially to differ from PPs, as when in 9.1.2 we replaced ρ_1 and ρ_2 by the combination of variables $(\rho_2 - \rho_1)$, or when in 5.4.1, having worked with a surface-tension component τ_z, we replaced this in the final result by $\tau \cos \theta$. But we were then supplementing dimensional considerations with conventional physical arguments and, as we have seen, the 'mixed' manner of approach does not necessarily confine itself merely to throwing up DPs.

Regarding this from another viewpoint, we may say that where the number of quantities exceeds the number of reference dimensions by only one, and where the situation is known to involve functions other than PPs, then the indicial matrix will inevitably reflect the situation by showing a singularity in any system that may be adopted. Thus, the current at time t during the discharge of a capacitor is given by

$$i = \frac{V}{R} e^{-t/RC}$$

Treating this in the MLT system, we have $(n - r) = (5 - 3) = 2$ DPs but clearly in $MLTQ$, or in any other system based on a greater number of reference dimensions than 3, the indicial matrix must be singular and it will be unreasonable of us to expect its rank to exceed 3, for then the number of DPs would reduce to only one and the solution would consequently be expressible as a PP — which is impossible Further examples of this type of situation were seen in 5.3.2 and 8.5.5.

The converse position does not hold, for we have frequently noticed that if there be 2 or more DPs there is no implication that the solution cannot be written as a single PP and we have, indeed, encountered many examples, particularly in 9.1, where a multiplicative combination of more than one DP finds a place in a complete solution. We have, then, to enquire whether or not an appropriate selection of reference dimensions and variables might not always enable us in such cases so to increase the rank of the matrix as to result in a direct determination of the one DP that does in fact comprise the solution.

11.4   Can a power-product relationship always be revealed?

Let us rephrase our question as follows: where a solution consists of a direct multiplicative relationship between powers, can dimensional analysis always be made to yield that solution, apart, that is, from the numerical coefficient?

In a (possibly) trivial sense the answer is 'yes' − provided that a sufficiently extended set of reference dimensions be used. Such a set will always be available, but its determination is liable to involve a certain sophistication and artificiality. That a suitable set exists may be demonstrated as follows. Let the solution be of the form $P = A^a B^b C^c \dots$ we have, then, merely to regard the independent variables $ABC \dots$ as themselves comprising the required reference set. The indicial matrix will then be

	A	B	$C \dots$
P	$-a$	$-b$	$-c$
A	1	0	0
B	0	1	0
C	0	0	1
.			
.			
.			

which has, of course, the required unique solution.

Similarly, the solution may be derived from any one of an infinite number of reference sets, each reference dimension being representable by a PP based on the quantities $ABC \dots$, provided always that the sets be so constructed as to satisfy the criteria of completeness. It is, however, in practice unlikely that we shall have the necessary knowledge to construct the required matrix, for, unless we know what is the form taken by the final result and unless we make use of the hindsight which will then be available, we shall be unable to express the dimensions of the dependent variable in terms of the particular reference set chosen. The fact is, however, that from a theoretical point of view we may always in principle reveal a PP by working in a sufficiently extended reference set.

Alternatively we may, again in principle, always reveal a PP by using as our variables sufficiently complex multiplicative groups based on the original 'primitive' quantities that enter into the situation. But we shall still require considerable sophistication if we are so to reduce the number of these complex variables that $(n − r)$ attains its limiting magnitude of unity. In such cases the reader will be at liberty to object that the solution has only been attained by resort to non-dimensional methods.

Nevertheless, in practice, the successful revelation of a 'difficult' PP will most generally be dependent upon the use of groups of variables that are brought to light as a result of our physical insight into the problem, although, as the reader will already have appreciated, reliance upon techniques leading both to a reduction in n and an increase in r may be used in the same problem. As we have seen before, both these approaches exemplify the principle that the greater the input of information the greater will be the precision of the result that is derived.

In illustration of this, we refer to the equation mentioned in 3.4 and expressing the energy release in an earthquake as $E = 8\pi^3 vtn^2 A^2 \rho d^2$. Here $n = 7$ and $r = 3$, implying that dimensional analysis would lead to $(7 - 3) = 4$ DPs. Nevertheless, the solution is itself clearly expressible as a single DP and we have to consider how best this may be revealed. We are unable to use the theoretically possible approach involved in using the quantities v, t, n, A, ρ and d as a reference set, for without knowing the solution we have no means of determining what would be the dimensions of the dependent variable E in this set. It seems, moreover, that no increase in the rank of the indicial matrix can be obtained by working in any more 'conventional' reference set. Nevertheless, we hesitate to abandon the dimensional approach and we attempt, therefore, to effect a suitable reduction in the number of variables by the use of judicious combinations. Note, then, that we have enquired directly, 'How does the total energy release depend upon the significant variables?' This may not, however, be the most heuristic approach. Let us instead ask, 'What is the energy content per unit volume of vibrating rock?' This second question may at once be answered by dimensional arguments, the solution being $E_v = k \cdot n^2 A^2 \rho$. But the total energy E is given by $E = E_v.V$, where V is the volume of the vibrating rock, a spherical shell of radius d and thickness vt. Dimensional considerations now show that $V = k.vtd^2$ and it follows that $E = k.vtd^2.n^2 A\rho$, which is equivalent to the desired result.

In commenting upon this, we see that the variables n, A and ρ are contained *only* in the composite variable E_v, while the variables v, t and d appear *only* in the composite variable V. Apart from these appearances, none of the variables n, A, ρ, v, t or d enters into the solution in its own right and, were one of them to do so, our approach would no longer have met with success (see again 9.1).

Consideration of this example suggests, then, that if we are to find the right answers, we must first ask the right questions. Admittedly, however, the right questions are often difficult to frame.

11.4.1 *Displacement due to Brownian motion*

The difficulty involved in ensuring that the variables used are sufficiently complex to reveal a PP relationship is well illustrated when we make an investigation into the Brownian displacement of a particle. The assumption usually made is that a particle in suspension has the same mean kinetic energy as would have a gas molecule at the same temperature. This suggests that the significant variables will be those listed in the table below:

Physical quantity	Symbol	M	F	L	T
Mean displacement of particle from origin	x	0	0	1	0
Time	t	0	0	0	1
Temperature	θ	0	1	1	0
Particle radius	r	0	0	1	0
Viscosity of fluid	μ	0	1	−2	1
Number of particles per unit mass	N	−1	0	0	0
Gas constant	R	−1	0	0	0

Dimensional analysis yields three DPs, namely (x/r), $(t\theta/x^3\mu)$ and (R/N). The solution, as derived from non-dimensional methods, however, is known to be simply

$$x^2 = \frac{R\theta}{N} \cdot \frac{t}{3\pi r\mu} \tag{1}$$

which is certainly a multiplicative relationship between powers of the quantities involved. We ask whether it may not be practicable to attain this result by dimensional methods.

Note that any attempt to extend the length dimension leads to no increase in the rank of the matrix, since in this problem component lengths cannot be regarded as physically independent. We have, therefore, to seek a reduction in the number of variables. Provided, now, that we have sufficient insight into the phenomenon, we may observe that

1. since viscosity is responsible for damping the particle movement, then r and μ will be combined to form the single quantity $(r\mu)$ as in Stokes' equation;

2. a term equivalent to the mean kinetic energy of the particle must enter the final result. But we have from the gas equation

$$\tfrac{1}{2}mv^2 = \frac{3}{2} \cdot \frac{R\theta}{N}$$

which implies that R, θ and N will be combined in the form $(R\theta/N)$ and this expression may again be treated as a single quantity.

Subject to these two restrictions, our three DPs reduce to one and we have

$$f\left(\frac{R\theta t}{x^2 N r \mu}\right) = 0$$

which is consistent with equation 1. Thus a somewhat complex DP has been revealed, but only by means of dimensional methods supplemented by arguments of an essentially physical nature.

This well illustrates the general position as we have stated it, for clearly in any given problem the solution of which consists of a PP, we may invariably attain that solution provided we have the sophistication to make sufficiently detailed restrictions upon the manner of combination of the primitive quantities concerned — that is, provided we are prepared to work with carefully chosen multiplicative groups derived from the original variables. To ask the right questions, then, is liable to involve ascertaining what those groups may be and subsequently enquiring what dimensional analysis may have to say as to the manner in which they may enter into the final solution. Admittedly, it may not be possible to determine the nature of the groups of quantities if dimensional analysis is to be our sole guide. In fact if we are to solve certain problems by dimensional methods, it may sometimes be necessary to have a shrewd idea of the answer before we set out on our task.

11.5   The attribution of a 'mystique' to dimensional analysis

There has been a tendency, particularly among British writers, to attribute something of a 'mystique' to dimensional analysis and to hold that dimensional formulae have almost esoteric implications in revealing the 'real' nature of the quantity concerned. It has been suggested, in effect, that the dimensional formula represents the physical identity of a quantity in a manner analogous to the way in which chemical formulae indicate the chemical identity of the corresponding compounds. Resulting from this idea, there has been much fruitless discussion as to which set of dimensions should be regarded as 'fundamental' or 'basic' and, once that set has been agreed upon, the dimensional formulae of quantities deriving from it tend to be regarded as, in some sense, absolute.

We have seen, however, that the dimensional representation of a quantity merely tells us the manner in which that quantity is measured and nothing as to its ultimate reality. Dingle[13] points out that there would, for instance, be no inconsistency in defining and measuring velocity in terms of the Doppler shift $d\lambda/\lambda$ (λ being a wavelength). Velocity would then be dimensionless, and we have as much right to say that velocity is 'really' dimensionless as to say that it 'really' has the dimensions of LT^{-1}. And we have already seen that it is a matter of indifference whether, using the MLT system, we represent force as having the dimensions MLT^{-2} or, using the FLT system, we represent mass as having the dimensions FT^2L^{-1}. Each approach is equally 'right' and, questions of convenience and tradition apart, there is no reason why we should opt for one rather than another.

Arising largely from the attitude that regards dimensions as pointers to the real nature of a quantity, there is a feeling that there *ought* to be a one-to-one correspondence between physical quantities and dimensional formulae. This is clearly not the case for, considering the MLT system, as we see from Appendix 1, heat, work and torque, for example are all equidimensional. Nevertheless, it may be instructive to consider the question further.

As a preliminary, we note that the more extensive the reference set, the less the probability of finding more than one quantity represented by the same formula. We may, for instance, readily distinguish between the dimensions of work and of torque as soon as we work with the $MXYZT$ system, for the definition of work involves the product of two colinear lengths and will be represented, say, by MX^2T^{-2}, while torque is defined in terms of the product of two mutually perpendicular lengths and will be represented, say, by $MXYT^{-2}$. Conversely, if we reduce the number of members of the reference set and if, in the limit, we work with one dimension only – a possibility mentioned in 9.4 – then there will generally be a multiplicity of physical quantities corresponding to each dimensional formula, provided always that the exponent is not impracticably large.

We ask, then, whether, in a sufficiently comprehensive reference system, there would be a unique formula available for each quantity; and conversely whether there is a unique physical quantity available to fit each formula.

In answering the first of these questions, we would reply 'yes', at least in a trivial sense, for, provided the situation is such that an indicial matrix may be constructed with linearly independent columns, then the two quantities which we wish to distinguish may each be regarded as a reference dimension in its own right (cf. the approach mentioned at the beginning of 11.4).

From a more realistic point of view, it is probable that pairs of complex and artificial quantities may usually be constructed with similar dimensional representations in any 'reasonable' reference set. But the fact remains that we are, nevertheless, certainly able to distinguish such pairs of naturally occurring quantities as angular velocity and frequency, or compressive strain and Poisson's ratio, in suitably extended sets, even though these be equidimensional in MLT.

Admittedly there is liable to be difficulty in distinguishing between ratios, for, no matter how far the reference set be extended, compressive strain, for instance, will be equidimensional with relative humidity unless some *ad hoc* artifice is introduced. Where quantities include a ratio in their definition, a distinction may again be difficult, as in the case of the pair of quantities Young's modulus and stress. This difficulty is also liable to arise with DPs, but generally a DP in one system will not necessarily be one in a sufficiently extended system. There is, incidentally, no contradiction here with the remark in **1.8** to the effect that a DP may always be represented as a ratio. Provided that in one system a combination of quantities is dimensionless, then it can always be represented as a ratio in the same system — but not necessarily in other systems.

We will not pursue this question further as it involves no real points of difficulty, but we briefly consider our second question concerning the possibility of associating a physical quantity with each arbitrary dimensional formula. There are, for instance, recognised quantities associated with LT^n for $n = 0, -1$ and -2, but what of the case where $n = +1$ or $+2$? And we are tempted to ask why, as may be seen from Table A1 of Appendix 1, the index of T tends to be -1 or -2, whereas that of M tends to be $+1$. Does any deep physical principle lie behind this tendency?

To clarify this latter point first, a little thought shows that the effect is largely a matter of convention. The negative index of T would be of less frequent occurrence if we were psychologically disposed to think in terms of 'slowness' rather than in terms of 'speed'. Similarly, there is no reason why we should not work more generally with 'lightness' or 'specific volume', defined as volume per unit mass, rather than with density. The adoption of a system based on these approaches would naturally result in a preponderance of *positive* indices for T and *negative* indices for M.

Admittedly there would be certain mathematical difficulties for, defining 'slowness' as dt/dx, a body at rest would have infinite slowness. And if, in this system, we define acceleration as 'change of slowness with distance', that is d^2t/dx^2, an instantaneous infinite acceleration would be required to set the body in motion. Nevertheless, a consistent and, indeed, elegant approach may be developed along these lines.

More generally, we may always define a quantity in such a way that it may be represented by the formula $M^a L^b T^c$ for arbitrary a, b and c, but how useful such a quantity may be is a matter of doubt. Thus, putting $a = 1, b = 2$ and $c = 3$, we could, were we so disposed, define a quantity $\psi = mr^2 t^3$ which might characterise a mass m rotating in a circular orbit of radius r in time t. We could then derive a theorem expressing the kinetic energy of the body by the equation $KE = 2\pi^2 \psi/t^5$. Other quantities could similarly be related to ψ and a plethora of trivial equations could be deduced. But although this would be perfectly permissible from a logical viewpoint, the introduction of artificial quantities corresponding to arbitrary dimensional formulae is unlikely to lead to any significant advance in our knowledge.

Little, then, is achieved by considering these approaches and most readers will agree that there is much to commend the attitude that refuses to see in dimensional analysis anything other than an enquiry into the logical implications of the manner in which we decide to define and measure such quantities as experience has suggested are of use in trying to understand the world of physical phenomena. Even so, these implications are of a surprising depth, and a sense of wonder at their range and scope is neither inappropriate nor unscientific.

11.6   Absolute units

We have said little or nothing concerning the choice of units for reference quantities — whether, for example, the metre is to be preferred to the yard, or the gram to the grain. From the viewpoint of the present book, this is a matter of little interest; but it must, nevertheless, be admitted that the units in current use are characterised by a certain arbitrariness. It is well to remind ourselves that even so 'scientific' a structure as the SI is based on the metre, defined originally in terms of the length of the earth's quadrant; the kilogram, defined originally in terms of the density of water; and the second, defined originally in terms of the rate of the earth's rotation about its axis.

In order to escape this element of arbitrariness, which the contemporary definitions set out in 1.3 do little to dispel, many have suggested that the use of an 'absolute' system of units should be adopted, a system in which the magnitude of the unit is so selected as to have a significance that transcends terrestrial conditions and reduces the value of certain important dimensional constants to unity. By so doing we should, as Bridgman[5] says, 'determine the size of the units by reference to universal phenomena instead of by reference to such restricted phenomena as the density of water at atmospheric pressure at some fixed temperature.'

Bridgman gives a number of examples of more rational approaches and mentions that Planck originally developed a system such that the gravitational, the gas and Planck's constants, together with the velocity of light, all reduce to unity. The resulting units of mass, length and time are then respectively $5\cdot43 \times 10^{-8}$ kg, $4\cdot02 \times 10^{-36}$ m and $1\cdot34 \times 10^{-43}$ seconds.

An alternative idea would be to define the unit of mass in terms, say, of the mass of the electron or proton; to define length in terms of the radius of the electron's orbit in the unexcited hydrogen atom; and to define time in terms of the same electron's orbital period. Such a system would again have claims to a universal rather than a merely local significance, but the practical advantages would be few.

In deriving a system of absolute units, we must bear in mind the requirements of 1.7, where we discussed the conditions for and the necessity of independence in a set of units. It follows from this that it would not be possible to construct a reference set of units from, say,

1. the velocity of light, Planck's constant and the charge on an electron, nor from

2. the gravitational constant, the charge on the electron and the mass of the electron,

for a brief investigation shows that both these sets of quantities are characterised by linear dependence.

It may be that a carefully chosen system of units will help simplify computation. Thus Eddington[19] occasionally found it helpful to work in a system in which the velocity of light and the product (Gh^2) were each taken as unity, G being the gravitational and h being Planck's constant. These two conditions left Eddington with one disposable unit which he generally took to be either mass or density.

11.7 Conclusion

In conclusion we would say that there are few physical problems that do not yield a partial solution when tackled by the techniques of dimensional analysis. This is so even when the problem is such that its complexity makes it impossible to give a precise formulation of the fundamental equations from which the mathematical solution is to be derived. Yet such phenomena yield as readily to a dimensional treatment as do ones in which the theoretical basis is well understood. This generality is at once the strength and the weakness of the dimensional method, for, while it has the widest range of application, it never reveals or even probes into the inner mechanism of a phenomenon. This inevitably remains hidden to the enquirer unless he is prepared to extend his investigation to include non-dimensional methods of approach.

11.8 Appendix on applications of dimensional analysis to cosmology

There are those who hold that further major developments in, for example, quantum theory and cosmology will be facilitated by the 'natural' expression of the quantities involved in dimensionless form rather than in terms of a largely arbitrary system of dimensional units. A number of cosmologists, for example, have been impressed by the frequency of unexpected numerical coincidences involving the force constant, defined as the ratio of the electric to the gravitational forces existing between proton and electron. This constant, e^2/Gm_pm_e, has the value $2 \cdot 3 \times 10^{39}$.

As an example, we assume with Milne and Lemaitre that the galaxies started from a common point at the time of creation $t = 0$ and in accordance with the celebrated 'big-bang' theory. Accepting Hubble's revised law of recession, the present age of the universe will be about 7×10^9 years. Note next that the time taken for light to cross the radius of an electron is about 10^{-23} seconds. Now the ratio of these two times is, of course, a pure number which increases proportionately with the age of the universe and which, at present, is equal to $2 \cdot 4 \times 10^{39}$ — which happens to be almost exactly the magnitude of the force constant defined above. Dirac[14] is one of those who consider that this connection is due to some fundamental link between cosmology and atomic theory.

Again, there is the cosmical number N obtained by dividing the total mass of the universe by the mass of a typical elementary particle. This has a magnitude of the order of 14×10^{78} which is approximately equal to the square of the force constant and, therefore, of the time ratio defined above.

By a series of apparent coincidences, it happens that many of the large dimensionless numbers of cosmology turn out to be of the order of either 10^{39} or 10^{78} and Dirac was sufficiently impressed by this to suggest that all such numbers of the order of 10^{39} are directly related to the time ratio, which he assumed to increase proportionately with the age of the universe. He further suggested that the magnitude of those constants of the order of 10^{78} increases with the square of the age. This principle was stated as:

> All very large dimensionless numbers which can be constructed from the important constants of cosmology and atomic theory are simple powers of the epoch (or age) with coefficients of the order of unity.

Other such numbers as $\rho c^3 / \lambda^3 m_p \approx 13 \times 10^{78}$ and $(hc/Gm_p{}^2)^2 = 1 \cdot 1 \times 10^{78}$ will be found listed by Schatzman[63]. (Here ρ is the mean density of the universe, $\lambda \equiv T^{-1}$ is Hubble's constant and h is Planck's constant.)

It would, for instance, follow from all this that the force constant itself increases directly with time, while the number of particles in the universe increases with the square of the time. This idea, however, does not carry a high order of conviction and if pursued further leads eventually to a number of difficulties which can only be resolved by making unrealistic and *ad hoc* assumptions. Indeed, one may be forgiven for suggesting that the argument has something of the flavour of the considerations that influenced the mediaeval numerologists.

Dimensional thinking also played a major role in Pascual Jordan's[37] approach to certain cosmological problems. He considered that there are six primary quantities which underly the structure of the universe, these being:

Physical quantity	Symbol	M	L	T	c.g.s value
Velocity of light	c	0	1	−1	3×10^{10}
Gravitational constant in general relativity = $8\pi G/c^2 =$	f	−1	1	0	2×10^{-27}
Age of universe	A	0	0	1	10^{17}
Mean density of universe	ρ	1	−3	0	10^{-28}
Hubble's constant of recession	λ	0	0	−1	10^{-17}
Radius of curvature of closed Riemannian space	R	0	1	0	3×10^{27}

From these quantities, $(6 - 3) = 3$ DPs may be formed, namely $(A\lambda)$, (R/cA) and $(f\rho c^2 A^2)$. Substituting the values given in the last column of the table gives

$$\lambda A = 1 \tag{1}$$

$$R/cA = 1 \tag{2}$$

$$f\rho c^2 A^2 = 1\cdot 8 \tag{3}$$

That each of these products should have a value approximating to unity suggests that there may be a real rather than a fortuitous relationship between the quantities entering into them (11.2). Supposing that this be the case, we differentiate equation 2 with respect to t to obtain

$$\frac{dR}{dt} = c\frac{dA}{dt} = c$$

which implies that the radius of the universe is increasing with the speed of light. Similarly, an implication of equation 1 is that it has been increasing with this speed since the occurrence of the 'big bang' at time $t = 0$ — as follows from the definition of Hubble's constant λ in terms of the equation $v = \lambda R$, where v is the velocity of recession at a distance R from the observer.

With regard to equation 3, we resort to some algebraic manipulation. Writing $\approx$ to denote equality to within the order of unity, we have

volume of universe $v \approx R^3$

mass of universe $m \approx R^3 \rho$

giving $\rho \approx m/R^3$

From equation 2, $cA \approx R$

Substituting these values for (cA) and ρ in equation 3 gives

$$f . m \approx R$$

 ...stituting the value of $8\pi G/c^2$ for f, as shown in the table, now gives

$$\frac{8\pi Gm^2}{R} \approx mc^2 \tag{4}$$

Each side of this final approximation now has the dimensions of $ML^2 T^{-2}$ and it follows from Einstein's equivalence relationship between mass and energy, $E = mc^2$, that the right-hand side of relationship 4 should be interpreted as the rest energy due to the total mass of the universe. The term of the left is the potential energy due to the gravitational forces consequent upon the distribution of this mass. It is, then, implied by Jordan's argument that there may well be an equality between these two quantities.

Another simple dimensional argument with a bearing upon the energy content of the universe is discussed by Gamow[28], who considers the density of radiation energy and the density of energy due to the presence of matter ($E = mc^2$). With regard to the former, Gamow shows that the mean temperature of an expanding universe decreases with its radius; and it follows from the Stefan–Boltzmann law, giving the energy radiated per unit volume as $E = \sigma\theta^4$, that the radiation density will be proportional to R^{-4}. The density of matter in the universe will, however, vary with R^{-3}.

Now, as a result of the very high temperatures existing in the small-radius universe during the early stages of its evolution, the density of radiant energy much initially have been greater than was the density of energy due to the presence of matter. It is, then, implied that with continued expansion this inequality must inevitably have become reversed, and Gamow calculates that the transition from the 'energetic' to the 'material' period occurred $2 \cdot 5 \times 10^8$ years after the initial 'big bang'.

Appendix 1

Dimensions of physical quantities

Table A1 Dimensions of some mechanical quantities expressed in terms of *MLT*

	Exponent of dimension					*Exponent of dimension*		
	M	*L*	*T*			*M*	*L*	*T*
Mass	1	0	0		Moment of inertia – mass	1	2	0
Length	0	1	0		Moment of inertia – area	0	4	0
Time	0	0	1		Surface tension	1	0	-2
Area	0	2	0		Dynamic viscosity	1	-1	-1
Volume	0	3	0		Kinematic viscosity	0	2	-1
Velocity	0	1	-1		Modulus of elasticity	1	-1	-2
Acceleration	0	1	-2		Strain	0	0	0
Density	1	-3	0		Compressibility	-1	1	2
Force	1	1	-2		Flexural rigidity	1	3	-2
Stress	1	-1	-2		Torsional rigidity	1	3	-2
Pressure	1	-1	-2		Poisson's ratio	0	0	0
Energy	1	2	-2		Efficiency	0	0	0
Work	1	2	-2		Weight	1	1	-2
Torque	1	2	-2		Angle	0	0	0
Moment of a force	1	2	-2		Solid angle	0	0	0
Power	1	2	-3		Frequency	0	0	-1
Momentum	1	1	-1		Angular velocity	0	0	-1
Impulse	1	1	-1		Angular acceleration	0	0	-2
Moment of momentum	1	2	-1		Curvature	0	-1	0
Angular momentum	1	2	-1					

Table A2 Dimensions of some thermal quantities expressed in various systems

Physical quantity	Symbol	M	L	T	Θ	M	L	T	H	M	L	T	H	Θ
Temperature	θ	0	0	0	1	0	0	0	0	0	0	0	0	1
Quantity of heat	q	1	2	-2	0	0	0	0	1	0	0	0	1	0
Specific heat capacity	c	0	2	-2	-1	-1	0	0	1	-1	0	0	1	-1
Thermal conductivity	κ	1	1	-3	-1	-1	-1	-1	1	-1	-1	-1	1	-1
Coefficient of heat transfer	h	1	0	-3	-1	-1	-2	-2	1	-1	-2	-2	1	-1
Entropy	s	1	2	-2	-1	-1	0	0	1	-1	0	0	1	-1
Enthalpy	H	1	2	-2	0	0	0	0	1	0	0	0	1	0
Internal energy	U	1	2	-2	0	0	0	0	1	0	0	0	1	0
Mechanical equivalent of heat	J	0	0	0	0	1	2	-2	-1	1	2	-2	-1	0
Gas constants	R, ℛ	0	2	-2	-1	-1	2	-2	0	-1	2	-2	0	-1
Coefficient of thermal expansion	β	0	0	0	-1	0	0	0	0	0	0	0	0	-1
Temperature gradient	dθ/dx	0	-1	0	1	0	-1	0	0	0	-1	0	0	1

Note: Specific quantities, defined in terms of unit mass, have the exponent of *M* reduced by one. For interrelationships between the systems used, see **6.2**.

Table A3 Dimensions of some electrical and magnetic quantities expressed in various systems

Physical quantity	Symbol	M	L	T	Q	M	L	T	μ	M	L	T	ε
Charge	q	0	0	0	1	½	½	0	−½	½	1½	−1	½
Current	i	0	0	−1	1	½	½	−1	−½	½	1½	−2	½
Potential difference	V	1	2	−2	−1	½	1½	−2	½	½	½	−1	−½
Resistance	R	1	2	−1	−2	0	1	−1	1	0	−1	1	−1
Inductance	L	1	2	0	−2	0	1	0	1	0	−1	2	−1
Capacitance	C	−1	−2	2	2	0	−1	2	−1	0	1	0	1
Reactance	X	1	2	−1	−2	0	1	−1	1	0	−1	1	−1
Impedance	Z	1	2	−1	−2	0	1	−1	1	0	−1	1	−1
Resistivity	ρ	1	3	−1	−2	0	2	−1	1	0	0	1	−1
Permittivity	ε	−1	−3	2	2	0	−2	2	−1	0	0	0	1
Electric field intensity	E	1	1	−2	−1	½	½	−2	½	½	−½	−1	−½
Electric displacement or induction	D	0	−2	0	1	½	−1½	0	−½	½	−½	−1	½
Magnetic pole	m	1	2	−1	−1	½	1½	−1	½	½	½	0	−½
Permeability	μ	1	1	0	−2	0	0	0	1	0	−2	2	−1
Magnetic field intensity	H	0	−1	−1	1	½	−½	−1	−½	½	½	−2	½
Magnetic induction	B	1	0	−1	−1	½	−½	−1	½	½	−1½	0	−½

Table A4 Dimensions of certain constants

Constant	Symbol	M	L	T	Θ	Units (SI)
Avogadro number	N	−1	0	0	0	mols/kmol
Boltzmann constant	k	1	2	−2	−1	J/K
Gravitational constant	G	−1	3	−2	0	Nm^2/kgm^2
Planck's constant	h	1	2	−1	0	J s
Radiation density constant	α	1	−1	−2	−4	$J/m^3 K^4$
Stefan−Boltzmann constant	σ	1	0	−3	−4	$J/m^2 s K^4$
Velocity of light	c	0	1	−1	0	m/s
Universal gas constant	$\mathcal{R}$	0	2	−2	−1	J/K

Appendix 2

Miscellaneous problems and points for discussion

1. What are the dimensions of pressure, viscosity and power in a system based on the reference dimensions density, velocity and length?

2. Show that if permittivity, permeability and the velocity of light were all considered dimensionless and of magnitude unity in free space, then resistance would also be a dimensionless quantity. What would be the magnitude of its unit?

3. Rework the conversion of lbf/in^2 into N/m^2 (1.6) using conversion factors based directly on the reference dimensions M, L and T.

4. In a particular situation, the DP A was found to depend upon the DPs B and C in the following manners:
   a) $A = \phi (B, C)$
   b) $A = \phi (BC)$
   c) $A = k . BC$

   What conclusions could be made about similarity in a model of the situation?

5. It is found empirically that over a certain velocity range the drag on a body is proportional to its cross-section raised to the power of its velocity: $D = k . A^v$. Discuss the manner in which this equation may be made dimensionally homogeneous and, in particular, how we may avoid a situation in which the dimensions of k vary with v. (Cf. a problem concerning a rheological equation discussed by Dingle[13].)

6. Show that the excess pressure in a soap bubble varies with the surface tension and inversely as the radius.

7. Show that the period of vibration of a rain drop, regarded as a surface-tension effect, is proportional to the radius raised to the power of 3/2 (Rayleigh).

8. Two tea leaves floating on the surface of a cup of tea are drawn together as a result of surface-tension forces. Investigate this effect dimensionally. How does the density of the liquid affect the situation? What is the final velocity of impact?

9. Show that the velocity of deepwater waves, dependent upon gravity, varies with the square root of (λg), while the velocity of surface ripples, dependent upon surface tension, varies with the square root of $(\tau/\rho\lambda)$. Discuss the intermediate case where both g and λ are significant.

10. In the manner of **9.6.2**, consider a jet in which the momentum flux, rather than the buoyancy flux, is dominant. Show that the dependence of the jet radius, velocity and concentration upon the distance from the source are given respectively by:

$$r = k.z$$
$$v = k.M^{1/2}z^{-1}$$
$$\sigma = k.\dot{m}M^{-1/2}z^{-1}$$

(M, here, is the momentum flux which, for the purposes of this problem, may be taken as $M = \pi r_0^2 u_0^2$.)

11. A 'comparative viscometer' is constructed from a tall cylindrical container with a hole punched in its base. The cylinder is filled with liquid and the time t taken for the level to fall between two marks is observed. Using water as a reference fluid, it is then found that $t = 21$ s. The cylinder is next filled with a liquid whose properties are to be investigated and it is found that $t = 93$ s. If the density of the fluid is 1200 kg/m^3, calculate its viscosity. (Tables show that at the ambient temperature of $15\,^\circ$C the viscosity of water is $1 \cdot 14$ Ns/m^2.)

12. Deduce Graham's law of diffusion to the effect that the velocity of gaseous molecules passing through a porous partition varies with the square root of the pressure difference across the partition and inversely as the square root of the gas density.

13. A cylindrical container is partially filled with a viscous liquid and is allowed to roll down an inclined plane. Investigate the possibility of determining the final velocity. Neglect the weight of the container.

14. With a minimum of physical reasoning, show that the force required to pull a body up a rough incline is proportional to $mg(\sin \alpha + k.\mu \cos \alpha)$, where α is the angle of the incline and μ is the coefficient of friction.

15. Consider the velocity of collapse v of a row of thin dominoes of height l and spacing d, each knocking down the next as it falls. Show that orthogonal directions are physically dependent and that we are, therefore, unable to obtain a complete solution to the problem by dimensional methods.

16. Assuming the orbits of planets are geometrically similar, find an expression
     for the length of year in terms of the relevant physical quantities, including
     the gravitational constant, and show that this varies with the length of the
     major axis of the orbit raised to the power of 3/2.

17. According to a once popular 'theory', gravitational force is due to particles
     which permeate space, moving with high velocities in random directions
     and impingeing upon material bodies in their path. The existence of such
     particles would certainly result in an attractive force between neighbouring
     bodies. Show that if the attraction between two similar spheres and due to
     such a mechanism is to fall off with the square of the distance between them,
     then it must also be proportional to the fourth power of their radius.

18. A model of a plane beam structure is made with axial lengths reduced by a
     factor s and cross-sectional areas reduced by a factor r. Show that for
     geometrically similar deflections between model and prototype we have
     $\Delta_m = s \cdot \Delta_p$. Show also that, provided bending and torsion only are significant,
     the model law is

$$f_m = \frac{r^4}{s^2} \cdot \frac{E_m}{E_p} \cdot f_p$$

     where f_m and f_p represent the corresponding characteristic forces in model
     and prototype. If the deflection is due solely to axial forces in the members,
     show that the model law will be:

$$f_m = r^2 \frac{E_m}{E_p} \cdot f_p$$

19. Show that the resolving power of an object glass, as measured by the
     reciprocal of the angle of the cone of light it can use, is directly
     proportional to its diameter and inversely proportional to the wavelength
     of the incident light.

20. Show that the force on a magnet due to the presence of another magnet
     varies as the product of the magnetic moments and inversely as the fourth
     power of the distance between them, this distance being assumed large
     compared with the lengths of the magnets.

21. Show that the square of the frequency of vibration of a magnetic needle
     in the plane of a uniform magnetic field is proportional to the intensity
     of the field and the magnetic moment of the magnet and inversely
     proportional to the moment of inertia of the needle about its axis of
     vibration.

22. The shaft power P of a d.c. motor is a function of the armature diameter d, the armature length l, the armature current i, the angular velocity of the shaft ω and the magnetic flux density B. Show that the relevant DPs are $(Bid^2\omega/P)$ and (l/d). Assuming that P is proportional to l, show that the complete solution to the problem is $P = k \cdot Bldi\omega$.

23. How would the frequency of a d.c.-operated electric buzzer be expected to vary with the applied voltage?

24. Discuss the difficulties liable to arise if we attempt to determine the velocity of compressive waves through an isotropic solid in terms of the extended $MXYZT$ set.

25. A partially submerged floating cylinder is so constrained that it oscillates with its axis vertical. Show that for a small oscillation $t = k \cdot (m/\rho Ag)^{1/2}$, where m is the mass of the cylinder, A its cross-sectional area and ρ the density of the liquid.

26. Show that the mass flow rate of a thin film of treacle moving down a flat vertical wall is given by

$$\dot{m} = k \cdot \rho^2 t^3 bg/\mu$$

where t is the thickness of the film, b is the width of the film, the other symbols having their usual meaning.

27. Show that for steady laminar natural convection of air up a vertical circular tube $Nu = \phi(Pr)$ and is independent of the Grashof number. (See 6.3.2.)

28. Discuss the possibility of representing enthalpy $H = U + pV$ in the $MLTH$ and $MLT\Theta$ systems.

29. The number of DPs in a complete set is given by $(n - r)$, where n is the number of variables and r is the rank of the indicial matrix. Show that the number of DPs is also given by the number of variables less the maximum number of variables that will not form a DP.

29. Prove that the number of DPs in a complete set is invariant with respect to the choice of reference dimensions, provided that these be independent.

30. Show, by resort to the 'dimension space' of 1.8 or otherwise, that if a PP is dimensionless with regard to one reference set, it will also be dimensionless with regard to any other set derived from the first by a linear transformation.

31. Why are the indices of a PP always rational? Would there be any objection to defining a quantity in such a way that its dimensional representation involves irrational indices?

32. Show that if the energy, E, emitted from a black body per unit area per unit time is dependent upon the absolute temperature θ, the velocity of light c. Planck's constant h and Boltzmann's constant k, then

$$E = k \cdot \frac{k^4}{c^2 h^3} \theta^4$$

(The dimensional representations of h and of k will be found in Table A4 of Appendix 1. The constant in this equation is shown analytically to be equal to $2\pi^5/15$.)

33. The energy of an electron in orbit is a function of its mass m, its charge e, the permittivity of the space between nucleus and electron ϵ and Planck's constant h. Show that

$$E = k \cdot \frac{me^4}{\epsilon h^2}$$

(Bohr's theory applied to a one-electron atom gives $E = 2\pi^2 \, me^4 Z^2/\epsilon h^2 n^2$, where (Ze) is the charge on the nucleus and where the nth permitted orbit is under consideration. Z and n are, therefore, both dimensionless.)

34. The frequency of radiation emitted when an electron passes from an energy state associated with orbit $n = j$ to a lower energy state associated with orbit $n = k$ is given by

$$f = 2\pi^2 \frac{me^4 Z^2}{\epsilon^2 h^3} \left(\frac{1}{k^2} - \frac{1}{j^2} \right)$$

Test this equation for dimensional homogeneity.

35. In plasma physics, the 'Debye shielding distance' λ, with dimension L, is dependent upon Boltzmann's constant k, the temperature of the electron gas θ, the number of ions and electrons per unit volume N and the charge on an electron. Show that

$$\lambda = \text{constant} \times \left[\frac{k\theta}{Ne^2} \right]^{1/2}$$

36. An electron is constrained to move along the axis of a ring charge. Show that the frequency of small amplitude oscillations will be given by:

$$\omega = k \cdot \left(\frac{eq}{\epsilon ma^2} \right)^{1/2}$$

where the total charge on the ring is q, the ring radius is a, the charge on the electron is e and the mass of the electron is m.

37. Milne suggested that ρ, the mean density of the universe, was related to its age A and the gravitational constant G. What would be the form of such a relationship? Discuss its plausibility given that

$\rho = 0.5 \times 10^{-25} \, \text{kg/m}^3$, $A = 10^{17}$ s and $G = 6.7 \times 10^{-11} \, \text{m}^3/\text{kg s}^2$.

References and Bibliography

In the following bibliography, works of general interest and not necessarily referred to in the text are marked with an asterisk.

1 ASIMOV, I. 'The light mile'. Article 108 in *The Scientist Speculates* (edited I. J. Good). Heinemann 1962
2 BOER, J. de. 'Quantum theory of condensed permanent gases' (together with a series of allied papers, some co-authored). *Physica* (The Hague) **XIV** 1948–9
3 BOND, W. N. *Phil. Mag.* **7** (94) 1929, 719–22
4 BRADSHAW, P. *An Introduction to Turbulence and its Measurement*. Pergamon 1971
*5 BRIDGMAN, P. W. *Dimensional Analysis*. Yale University Press. Revised edition 1963
*6 BRIDGMAN, P. W. 'Dimensional analysis'. *Encyclopaedia Brittanica*. 1963 edition
7 BROWN, G. Burniston. *Proc. Phys. Soc.* **53**(4) 1941, 418–32
8 BUCKINGHAM, E. 'On physically similar systems: illustrations of the use of dimensional equations'. *Phys. Rev.* **IV**(4) 1914, 345–76
9 CARR, L. H. A. 'The MKS or Giorgi system of units'. *Proc. IEE.* Vol **97** (I) 1954, no. 107
10 CORRSIN, S. 'A simple geometrical proof of Buckingham's π-theorem'. *Am. J. Phys.* **19** 1951, 180–1
11 COSTA, F. V. 'Directional analysis in model design'. *Proc. ASCE J. Eng. Mech. Div.* April 1971
12 DINGLE, H. *Phil. Mag.* **33** (220) 1942, 327–30
13 DINGLE, H. *Phil. Mag.* **40** (300) 1949, 94–9
14 DIRAC, P. A. M. *Nature.* 139:323. 20 Feb. 1937
*15 DUNCAN, W. J. *Physical Similarity and Dimensional Analysis*. Edward Arnold 1953
16 DUNCAN, W. J. THOM, A. S. and YOUNG, A. D. *Mechanics of Fluids*. Edward Arnold, 2nd edition 1970
17 DUNCANSON, W. E. *Proc. Phys. Soc.* **53**(4) 1941, 432–48
18 EDDINGTON, A. S. *The Philosophy of Physical Science*. Cambridge University Press 1939
19 EDDINGTON, A. S. *Proc. Roy. Soc. (A).* **174**(18) 1940
20 EINSTEIN, A. *Annales der Physik.* **35** 1911, 686
21 EISNER, F. Das Widerstandproblem. *Proc. 3rd Intern. Congr. Appl. Mech.* Stockholm 1931

*22 ESNAULT–PELTERIE, R. *L'Analyse Dimensionelle*. Rouge (Lausanne) 1946
 23 FALKNER, V. M. and SKAN, W. S. 'Some approximate solutions of the boundary layer equations'. *Phil. Mag.* **12** 1931, 865
 24 FERRAR, W. L. *Algebra*. Oxford 1941
*25 FOCKEN, C. M. *Dimensional Methods and Their Applications*. Edward Arnold 1953
 26 FORD, K. W. 'Magnetic monopoles'. *Scientific American*. **209**(6) December 1963, 123–31
 27 GALILEI, GALILEO. *Dialogues Concerning Two New Sciences*. 1634
 28 GAMOW, G. *The Creation of the Universe*. Viking Press (New York) 1952
 29 GESSLER, JOHANNES. 'Vectors in dimensional analysis'. *Proc. ASCE J. Eng. Mech. Div.* **99**(EM1) Feb. 1973, 121–9
 30 GOLDSTEIN, S. 'A note on the boundary layer equations'. *Proc. Camb. Phil. Soc.* **35** 1939, 338–40
 31 HAINZL, J. 'On local generalisations of the pi theorem of dimensional analysis'. *J. Frank. Inst.* **292**(6) 1971, 463–70
 32 HAMILTON, J. H., KUSIAN R. N., and COPE W. J., 'Fan performance and selection using dimensionless ratios'. *Bull. Univ. Utah.* **45**(12) 1954
 33 HAPP, W. W. 'Dimensional analysis via directed graphs.' *J. Frank. Inst.* **292**(6) 1971, 527–33
*34 HUNTLEY, H. E. *Dimensional Analysis*. Macdonald 1952
*35 IPSEN, E. C. *Units, Dimensions and Dimensionless Numbers*. McGraw-Hill 1960
 36 JOOS, GEORG. *Theoretical Physics*. Blackie, 3rd edition 1958
 37 JORDAN, PASCUAL. *Ann. der Phys.* **36** 1939, 64
*38 JOURNAL OF THE FRANKLIN INSTITUTE. *Modern Dimensional Analysis, Similitude and Similarity*. **292** (6) Pergamon Press 1971
*39 JUPP, E. E. *An Introduction to Dimensional Methods*. Cleaver-Hume 1962
*40 LANCHESTER, F. M. *The Theory of Dimensions and its Applications for Engineers*. Crosby Lockwood 1936
*41 LANGHAAR, HENRY L. *Dimensional Analysis and the Theory of Models*. Chapman and Hall 1951
 42 LIEPMANN, H. W. and ROSHKO, A. *Elements of Gasdynamics*. Wiley 1957
 43 LITTLEWOOD, D. E. *A University Algebra*. Heinemann 1950
 44 LITTLEWOOD, J. E. *A Mathematician's Miscellany*. Methuen 1953
 45 *Encyclopaedia of Science and Technology*. McGraw-Hill, 3rd edition 1970
*46 MASSEY, B. S. *Units, Dimensional Analysis and Physical Similarity*. Van Nostrand Reinhold 1971
 47 MAXWELL, J. CLERK. 'On the mathematical classification of physical quantities'. *Proc. Lond. Math. Soc.* **III**(34) March 1871, 224 et seq.
 48 MAXWELL, J. CLERK. *Treatise on Electricity and Magnetism*. Oxford 1892
 49 MONOD, JAQUES. *Chance and Necessity*. Collins 1972
 50 MORAN, M. J. 'A generalization of dimensional analysis'. *J. Frank. Inst.* **292**(6) Dec. 1971, 423–32
 51 NA, TSUNG Y. & HANSEN, ARTHUR G. 'Similarity analysis of differential equations by Lie group'. *J. Frank. Inst.* **292**(6) Dec. 1971, 471–89
 52 PANKHURST, R. C. 'Alternative formulation of the pi-theorem'. *J. Frank. Inst.* **292**(6) Dec. 1971, 451–62
 53 PEARSON, K. *Grammar of Science*. A & C Black 1900 (also Dent Everyman edition 1937)

54 PEARSONS, OLIVER T. 'Metabolism of humming-birds'. *Scientific American.* **188** 1953, 69–72
55 PORTER, ALFRED W. *The Method of Dimensions.* Methuen 1933
56 RAYLEIGH, (Lord) J. W. S. *Theory of Sound,* I, Art. 163, 1877–8 (Dover edition 1945)
57 RAYLEIGH, (Lord) J. W. S. 'On the viscosity of argon as affected by temperature'. *Proc. Roy. Soc. Lond.* **LXVI** 1899–1900, 68–74
58 RAYLEIGH, (Lord) J. W. S. *Nature.* 95:2368:67 1915
59 RIABOUCHINSKY, D. *Nature.* 95:2387:591 1915
60 RUCKER, Sir A. W. *Phil. Mag.* 27:165, 1889, 104–14
61 RUTHERFORD, D. E. *Fluid Dynamics.* Oliver and Boyd 1959
62 SANDON, H. *Essays in Protozoology.* Hutchinson 1959
63 SCHATZMAN, EVRY. *The Origin and Evolution of the Universe.* Hutchinson 1966
64 SCHLICHTING, H. *Boundary Layer Theory.* 1st English edition Pergamon Press 1955
65 STAICU, C. I. 'General dimensional analysis'. *J. Frank. Inst.* **292**(6) 1971, 433–9
*66 SEDOV, L. I. *Similarity and Dimensional Methods in Mechanics.* Academic Press (New York) 1959
*67 STUBBINGS, G. W. *Dimensions in Engineering Theory.* Crosby Lockwood 1948
68 THOMPSON, D'ARCY W. *On Growth and Form.* Cambridge 1948
69 TURNER, J. S. *Buoyancy Effects in Fluids.* Cambridge 1973
70 *Van Nostrand Scientific Encyclopaedia.* 3rd edition 1958
71 VON KARMAN, T. *Proc. 3rd Congr. Appl. Mech.* Stockholm 1930, 85–93
72 YALIN, M. S. *Theory of Hydraulic Models.* Macmillan 1971

Index